VOLUME THREE HUNDRED AND SEVENTY

INTERNATIONAL REVIEW OF CELL AND MOLECULAR BIOLOGY

Adoptive Cell Transfer

INTERNATIONAL REVIEW OF CELL AND MOLECULAR BIOLOGY

Series Editors

GEOFFREY H. BOURNE	*1949–1988*
JAMES F. DANIELLI	*1949–1984*
KWANG W. JEON	*1967–2016*
MARTIN FRIEDLANDER	*1984–1992*
JONATHAN JARVIK	*1993–1995*
LORENZO GALLUZZI	*2016–*

Editorial Advisory Board

AARON CIECHANOVER	WALLACE MARSHALL
SANDRA DEMARIA	SHIGEKAZU NAGATA
SILVIA FINNEMANN	MOSHE OREN
KWANG JEON	ANNE SIMONSEN
CARLOS LOPEZ-OTIN	

VOLUME THREE HUNDRED AND SEVENTY

INTERNATIONAL REVIEW OF CELL AND MOLECULAR BIOLOGY
Adoptive Cell Transfer

Edited by

FERNANDO ARANDA
Cima Universidad de Navarra, Pamplona, Spain

PEDRO BERRAONDO
Cima Universidad de Navarra, Pamplona, Spain

LORENZO GALLUZZI
*Weill Cornell Medical College,
New York, NY, United States*

Academic Press is an imprint of Elsevier
50 Hampshire Street, 5th Floor, Cambridge, MA 02139, United States
525 B Street, Suite 1650, San Diego, CA 92101, United States
The Boulevard, Langford Lane, Kidlington, Oxford OX5 1GB, United Kingdom
125 London Wall, London, EC2Y 5AS, United Kingdom

First edition 2022

Copyright © 2022 Elsevier Inc. All rights reserved.

No part of this publication may be reproduced or transmitted in any form or by any means, electronic or mechanical, including photocopying, recording, or any information storage and retrieval system, without permission in writing from the publisher. Details on how to seek permission, further information about the Publisher's permissions policies and our arrangements with organizations such as the Copyright Clearance Center and the Copyright Licensing Agency, can be found at our website: www.elsevier.com/permissions.

This book and the individual contributions contained in it are protected under copyright by the Publisher (other than as may be noted herein).

Notices
Knowledge and best practice in this field are constantly changing. As new research and experience broaden our understanding, changes in research methods, professional practices, or medical treatment may become necessary.

Practitioners and researchers must always rely on their own experience and knowledge in evaluating and using any information, methods, compounds, or experiments described herein. In using such information or methods they should be mindful of their own safety and the safety of others, including parties for whom they have a professional responsibility.

To the fullest extent of the law, neither the Publisher nor the authors, contributors, or editors, assume any liability for any injury and/or damage to persons or property as a matter of products liability, negligence or otherwise, or from any use or operation of any methods, products, instructions, or ideas contained in the material herein.

ISBN: 978-0-323-99399-9
ISSN: 1937-6448

For information on all Academic Press publications
visit our website at https://www.elsevier.com/books-and-journals

Publisher: Zoe Kruze
Acquisitions Editor: Leticia M. Lima
Developmental Editor: Jhon Michael Peñano
Production Project Manager: Vijayaraj Purushothaman
Cover Designer: Matthew Limbert

  Working together to grow libraries in developing countries

www.elsevier.com • www.bookaid.org

Typeset by STRAIVE, India

Contents

Contributors ix

1. **Impact of tumor microenvironment on adoptive T cell transfer activity**   1
   Celia Martín-Otal, Flor Navarro, Noelia Casares, Aritz Lasarte-Cía, Inés Sánchez-Moreno, Sandra Hervás-Stubbs, Teresa Lozano, and Juan José Lasarte

   1. Introduction   2
   2. T cell homing   3
   3. Physical barriers and the extracellular matrix   7
   4. Tumor infiltrating immunosuppressor cells   10
   5. Metabolic competition and interstitial fluid   15
   6. Concluding remarks   20
   Acknowledgments   21
   References   21

2. **Dendritic cell transfer for cancer immunotherapy**   33
   Liwei Zhao, Shuai Zhang, Oliver Kepp, Guido Kroemer, and Peng Liu

   1. Introduction   34
   2. Morphological and functional characteristics of DC   35
   3. Role of DCs in cancer immunotherapy   36
   4. Concept of DC-based immunotherapies   40
   5. Experimental and clinical approaches to improve DC-based immunotherapies   46
   6. Engineering dendritic cells to improve DC-based cancer immunotherapy   51
   7. Concluding remarks   52
   Acknowledgments   54
   Conflict of interest   54
   References   54

3. **Killers on the loose: Immunotherapeutic strategies to improve NK cell-based therapy for cancer treatment**   65
   Cordelia Dunai, Erik Ames, Maria C. Ochoa, Myriam Fernandez-Sendin, Ignacio Melero, Federico Simonetta, Jeanette Baker, and Maite Alvarez

   1. Introduction   66
   2. Immunomodulatory strategies   69

3. NK-based cellular therapy	83
4. NK immune cell engagers	90
5. Oncolytic virotherapy	93
6. Conclusions	97
Acknowledgments	99
Conflict of interest	99
References	99

4. Enabling CAR-T cells for solid tumors: Rage against the suppressive tumor microenvironment — 123
Asier Antoñana-Vildosola, Samanta Romina Zanetti, and Asis Palazon

1. Introduction	124
2. CAR-T cells targeting the tumor vasculature	127
3. CAR-T cells targeting cancer-associated fibroblasts	129
4. CAR-T cells targeting tumor associated macrophages and myeloid suppressor cells	131
5. Concluding remarks	136
Acknowledgments	136
References	136

5. What will (and should) be improved in CAR immunotherapy? — 149
Europa Azucena González-Navarro, Marta Español, Natalia Egri, Maria Castellà, Hugo Calderón, Carolina España, Carla Guijarro, Libertad Heredia, Mariona Pascal, and Manel Juan Otero

1. General concepts	150
2. Detection of Targets: Specificity and Affinity	151
3. Global molecular structure: From signaling to combination	153
4. Gene transfer protocols: Vectors, gene editing and upper CAR generation	154
5. Cell involvement: Allogenic, autologous and cell populations	155
6. Clinical protocols for personalized use of the product	157
7. Combination of therapies and procedures	158
8. Regulatory improvements	158
9. Pharma and Academic collaboration for supplying sustainable options to the health systems	159
10. Final discussion and conclusion	159
Acknowledgments	160
References	160

6. Adoptive tumor infiltrating lymphocyte transfer as personalized immunotherapy — 163
Ines Diaz-Cano, Luis Paz-Ares, and Itziar Otano

1. Introduction	164
2. History of TILs	165
3. Clinical studies	166
4. TIL manufacture	170
5. Antigen-specificity of TILs	172
6. T-cell differentiation state of TILs	178
7. Combinatorial therapies	181
8. Synthetic TILs in cancer therapy	182
9. Perspectives and conclusions	184
Acknowledgments	185
Author contributions	185
Competing interests	185
References	185

Contributors

Maite Alvarez
Program for Immunology and Immunotherapy, CIMA, Universidad de Navarra; Navarra Institute for Health Research (IdiSNA), Pamplona; Centro de Investigación Biomédica en Red de Cáncer (CIBERONC), Madrid, Spain

Erik Ames
Department of Pathology, Stanford University, Stanford, CA, United States

Asier Antoñana-Vildosola
Cancer Immunology and Immunotherapy Lab, CIC bioGUNE, Basque Research and Technology Alliance (BRTA), Bizkaia, Spain

Jeanette Baker
Blood and Marrow Transplantation, Stanford University School of Medicine, Stanford, CA, United States

Hugo Calderón
Servei d'Immunologia. Hospital Clínic de Barcelona. IDIBAPS. Universitat de Barcelona. Plataforma d'Immunoteràpia Hospital Clínic—Hospital Sant Joan de Déu, Barcelona, Spain

Noelia Casares
Program of Immunology and Immunotherapy, Cima Universidad de Navarra; Navarra Institute for Health Research (IdiSNA), Pamplona, Spain

Maria Castellà
Servei d'Immunologia. Hospital Clínic de Barcelona. IDIBAPS. Universitat de Barcelona. Plataforma d'Immunoteràpia Hospital Clínic—Hospital Sant Joan de Déu, Barcelona, Spain

Ines Diaz-Cano
H12O-CNIO Lung Cancer Clinical Research Unit, Health Research Institute Hospital 12 de Octubre/Spanish National Cancer Research Center (CNIO), Madrid, Spain

Cordelia Dunai
Department of Clinical Infection, Microbiology and Immunology, University of Liverpool, Liverpool, United Kingdom

Natalia Egri
Servei d'Immunologia. Hospital Clínic de Barcelona. IDIBAPS. Universitat de Barcelona. Plataforma d'Immunoteràpia Hospital Clínic—Hospital Sant Joan de Déu, Barcelona, Spain

Carolina España
Servei d'Immunologia. Hospital Clínic de Barcelona. IDIBAPS. Universitat de Barcelona. Plataforma d'Immunoteràpia Hospital Clínic—Hospital Sant Joan de Déu, Barcelona, Spain

Marta Español
Servei d'Immunologia. Hospital Clínic de Barcelona. IDIBAPS. Universitat de Barcelona. Plataforma d'Immunoteràpia Hospital Clínic—Hospital Sant Joan de Déu, Barcelona, Spain

Myriam Fernandez-Sendin
Program for Immunology and Immunotherapy, CIMA, Universidad de Navarra; Navarra Institute for Health Research (IdiSNA), Pamplona, Spain

Europa Azucena González-Navarro
Servei d'Immunologia. Hospital Clínic de Barcelona. IDIBAPS. Universitat de Barcelona. Plataforma d'Immunoteràpia Hospital Clínic—Hospital Sant Joan de Déu, Barcelona, Spain

Carla Guijarro
Servei d'Immunologia. Hospital Clínic de Barcelona. IDIBAPS. Universitat de Barcelona. Plataforma d'Immunoteràpia Hospital Clínic—Hospital Sant Joan de Déu, Barcelona, Spain

Libertad Heredia
Servei d'Immunologia. Hospital Clínic de Barcelona. IDIBAPS. Universitat de Barcelona. Plataforma d'Immunoteràpia Hospital Clínic—Hospital Sant Joan de Déu, Barcelona, Spain

Sandra Hervás-Stubbs
Program of Immunology and Immunotherapy, Cima Universidad de Navarra; Navarra Institute for Health Research (IdiSNA), Pamplona, Spain

Manel Juan Otero
Servei d'Immunologia. Hospital Clínic de Barcelona. IDIBAPS. Universitat de Barcelona. Plataforma d'Immunoteràpia Hospital Clínic—Hospital Sant Joan de Déu, Barcelona, Spain

Oliver Kepp
Metabolomics and Cell Biology Platforms, Gustave Roussy Cancer Center, Université Paris Saclay, Villejuif; Centre de Recherche des Cordeliers, Equipe labellisée par la Ligue contre le cancer, Université de Paris, Sorbonne Université, Inserm U1138, Institut Universitaire de France, Paris, France

Guido Kroemer
Metabolomics and Cell Biology Platforms, Gustave Roussy Cancer Center, Université Paris Saclay, Villejuif; Centre de Recherche des Cordeliers, Equipe labellisée par la Ligue contre le cancer, Université de Paris, Sorbonne Université, Inserm U1138, Institut Universitaire de France; Institut du Cancer Paris Carpem, Department of Biology, Hôpital Européen Georges Pompidou, APHP, Paris, France

Juan José Lasarte
Program of Immunology and Immunotherapy, Cima Universidad de Navarra; Navarra Institute for Health Research (IdiSNA), Pamplona, Spain

Aritz Lasarte-Cía
Program of Immunology and Immunotherapy, Cima Universidad de Navarra, Pamplona, Spain

Peng Liu
Metabolomics and Cell Biology Platforms, Gustave Roussy Cancer Center, Université Paris Saclay, Villejuif; Centre de Recherche des Cordeliers, Equipe labellisée par la Ligue contre le cancer, Université de Paris, Sorbonne Université, Inserm U1138, Institut Universitaire de France, Paris, France

Teresa Lozano
Program of Immunology and Immunotherapy, Cima Universidad de Navarra, Pamplona, Spain

Celia Martín-Otal
Program of Immunology and Immunotherapy, Cima Universidad de Navarra, Pamplona, Spain

Ignacio Melero
Program for Immunology and Immunotherapy, CIMA, Universidad de Navarra; Navarra Institute for Health Research (IdiSNA); Department of Immunology and Immunotherapy, Clínica Universidad de Navarra, Pamplona; Centro de Investigación Biomédica en Red de Cáncer (CIBERONC), Madrid, Spain

Flor Navarro
Program of Immunology and Immunotherapy, Cima Universidad de Navarra, Pamplona, Spain

Maria C. Ochoa
Program for Immunology and Immunotherapy, CIMA, Universidad de Navarra; Navarra Institute for Health Research (IdiSNA), Pamplona; Centro de Investigación Biomédica en Red de Cáncer (CIBERONC), Madrid, Spain

Itziar Otano
H12O-CNIO Lung Cancer Clinical Research Unit, Health Research Institute Hospital 12 de Octubre/Spanish National Cancer Research Center (CNIO); Spanish Center for Biomedical Research Network in Oncology (CIBERONC), Madrid, Spain

Asis Palazon
Cancer Immunology and Immunotherapy Lab, CIC bioGUNE, Basque Research and Technology Alliance (BRTA); Ikerbasque, Basque Foundation for Science, Bizkaia, Spain

Mariona Pascal
Servei d'Immunologia. Hospital Clínic de Barcelona. IDIBAPS. Universitat de Barcelona. Plataforma d'Immunoteràpia Hospital Clínic—Hospital Sant Joan de Déu, Barcelona, Spain

Luis Paz-Ares
H12O-CNIO Lung Cancer Clinical Research Unit, Health Research Institute Hospital 12 de Octubre/Spanish National Cancer Research Center (CNIO); Spanish Center for Biomedical Research Network in Oncology (CIBERONC); Medicine and Physiology Department, School of Medicine, Complutense University of Madrid, Madrid, Spain

Inés Sánchez-Moreno
Program of Immunology and Immunotherapy, Cima Universidad de Navarra, Pamplona, Spain

Federico Simonetta
Division of Hematology, Department of Oncology, Geneva University Hospitals; Translational Research Centre in Onco-Haematology, Faculty of Medicine, Department of Pathology and Immunology, University of Geneva, Geneva, Switzerland

Samanta Romina Zanetti
Cancer Immunology and Immunotherapy Lab, CIC bioGUNE, Basque Research and Technology Alliance (BRTA), Bizkaia, Spain

Shuai Zhang
Metabolomics and Cell Biology Platforms, Gustave Roussy Cancer Center, Université Paris Saclay, Villejuif; Centre de Recherche des Cordeliers, Equipe labellisée par la Ligue contre le cancer, Université de Paris, Sorbonne Université, Inserm U1138, Institut Universitaire de France, Paris, France

Liwei Zhao
Metabolomics and Cell Biology Platforms, Gustave Roussy Cancer Center, Université Paris Saclay, Villejuif; Centre de Recherche des Cordeliers, Equipe labellisée par la Ligue contre le cancer, Université de Paris, Sorbonne Université, Inserm U1138, Institut Universitaire de France, Paris, France

CHAPTER ONE

Impact of tumor microenvironment on adoptive T cell transfer activity

Celia Martín-Otal[a,†], Flor Navarro[a,†], Noelia Casares[a,b], Aritz Lasarte-Cía[a], Inés Sánchez-Moreno[a], Sandra Hervás-Stubbs[a,b], Teresa Lozano[a,*,‡], and Juan José Lasarte[a,b,*,‡]

[a]Program of Immunology and Immunotherapy, Cima Universidad de Navarra, Pamplona, Spain
[b]Navarra Institute for Health Research (IdiSNA), Pamplona, Spain
*Corresponding author: e-mail address: tlmoreda@unav.es; jjlasarte@unav.es

Contents

1. Introduction	2
2. T cell homing	3
3. Physical barriers and the extracellular matrix	7
4. Tumor infiltrating immunosuppressor cells	10
5. Metabolic competition and interstitial fluid	15
6. Concluding remarks	20
Acknowledgments	21
References	21

Abstract

Recent advances in immunotherapy have revolutionized the treatment of cancer. The use of adoptive cell therapies (ACT) such as those based on tumor infiltrating lymphocytes (TILs) or genetically modified cells (transgenic TCR lymphocytes or CAR-T cells), has shown impressive results in the treatment of several types of cancers. However, cancer cells can exploit mechanisms to escape from immunosurveillance resulting in many patients not responding to these therapies or respond only transiently.

The failure of immunotherapy to achieve long-term tumor control is multifactorial. On the one hand, only a limited percentage of the transferred lymphocytes is capable of circulating through the bloodstream, interacting and crossing the tumor endothelium to infiltrate the tumor. Metabolic competition, excessive glucose consumption, the high level of lactic acid secretion and the extracellular pH acidification, the shortage of essential amino acids, the hypoxic conditions or the accumulation of fatty acids in the tumor

---

[†] These authors contributed equally to this work.

[‡] These authors share senior co-authorship.

microenvironment (TME), greatly hinder the anti-tumor activity of the immune cells in ACT therapy strategies. Therefore, there is a new trend in immunotherapy research that seeks to unravel the fundamental biology that underpins the response to therapy and identifies new approaches to better amplify the efficacy of immunotherapies. In this review we address important aspects that may significantly affect the efficacy of ACT, indicating also the therapeutic alternatives that are currently being implemented to overcome these drawbacks.

1. Introduction

The failure of immunotherapy to achieve long-term tumor control is multifactorial. One of the main problems of adoptive T cell therapy (ACT) is the low efficiency of the injected cells to infiltrate the tumor. The transferred lymphocyte must find the tumor while circulating through the bloodstream but unfortunately, only a small fraction of the large number of transferred T cells eventually reach the tumor (Bernhard et al., 2008; Hong et al., 2011; John et al., 2013; Pockaj et al., 1994). It is well known that one of the main mediators in lymphocyte trafficking is inflammation. However, its protumorogenic effects on cells and its effects on the tumor endothelium and its stroma greatly hinder the transit of lymphocytes into the tumor to interact with cancer cells. The production of enzymatic and lipid mediators, chemokines and cytokines affect T cell trafficking, and in many cases, the balance between the recruitment of pro-tumor or anti-tumor cells is inclined toward the former, favoring tumor escape from immunosurveillance.

The tumor vasculature constitutes in many cases the so-called tumor endothelial barrier inhibiting the entrance of effector T cells (Castermans and Griffioen, 2007). Migration of lymphocytes into tumor tissues is a multistep process involving rolling, activation, adhesion and transendothelial migration (Lawrence et al., 1997). The transferred lymphocytes have to interact first with the tumor endothelium through interactions between mucins, selectins, and adhesion molecules (Adams and Shaw, 1994; Hogg et al., 2011). Structural abnormalities in tumor vasculature and the underlying stroma lead to tumor vessels lacking the normal monolayer of endothelial cells and this combined with abnormal pericyte coverage greatly alters the transendothelial migration of immune system cells. On the other hand, insufficient blood flow results in a hypoxic microenvironment and cancer cell metabolic shift from oxidative phosphorylation to glycolysis (the Warburg effect) (Vaupel and Hockel, 2000). Glucose consumption, extracellular pH (pH_e) acidification, the poverty in essential amino acids or the accumulation of fatty acids make the tumor microenvironment (TME) a hostile setting for the action of antitumor T lymphocytes. The discovery of these immunosuppressive mechanisms posed by the TME has highlighted the importance of a deeper understanding of the mechanism governing T cell trafficking, activation, exhaustion or unresponsiveness as well as the tumor immune evasion strategies. Here, we address the important aspects that significantly affect the efficacy of ACT therapies (summarized in a graphical abstract, Fig. 1), also briefly describing some of the strategies that are emerging in the preclinical and clinical setting to overcome these obstacles (Fig. 2).

2. T cell homing

Tumors disrupt and thwart T cell infiltration through tumor-directed aberrancies of endothelial vessels and adhesion molecule expression, chemokine-chemokine receptor mismatching, immunosuppression, and recruitment of cancer-associated fibroblasts (Bellone and Calcinotto, 2013). Immune cell infiltration into solid tumors and their movement within the TME are controlled by gradients of chemokines that interact with specific chemokine receptors in a process called chemotaxis. Dysregulated chemokine signaling in the TME favors tumor growth, the exclusion of effector immune cells, and the recruitment of immunosuppressive cells (Kohli et al., 2022). Understanding the chemotactic environment of solid tumors and identifying chemokines that regulate immune cell entry into solid tumors is a necessary step for the improvement of current immunotherapeutic interventions, including ACT therapies or immune checkpoint blockade.

Upon activation, T cells upregulate the expression of CXCR3 to respond to the interferon (IFN) inducible chemokines CXCL9, CXCL10 and CXCL11. Expression of these chemokines favors effector T cell infiltration into the tumor (Chheda et al., 2016; Hensbergen et al., 2005; Mikucki et al., 2015). In fact, these interferon-inducible CXC-chemokines are crucial immune modulators and survival predictors in colorectal cancer (Kistner et al., 2017). Intratumoral injection of IFNγ has been shown to increase the level of these chemokines in sarcoma or melanoma patients favoring T cell infiltration (Mauldin et al., 2016; Zhang et al., 2019). Direct strategies to increase the level of CXCR3 ligands in the tumor either by using viral vectors or by intratumor injection of the protein have been evaluated in preclinical models (reviewed in Kohli et al., 2022) even in combination with ACT therapies (Wang et al., 2013a). Furthermore, CXCR4

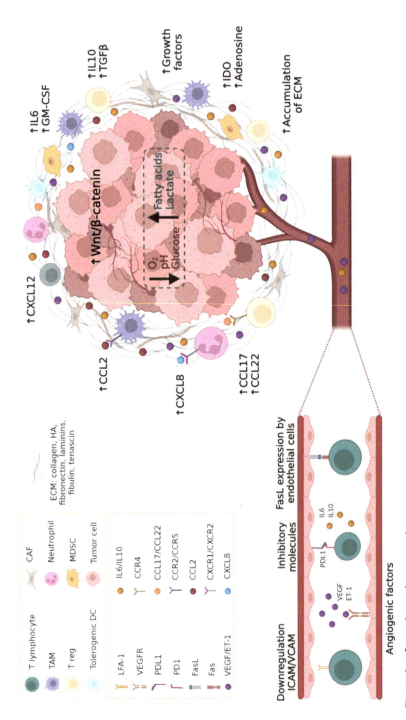

Fig. 1 See figure legend on opposite page.

expressing T cells respond to CXCL12 produced in the tumor. However, the CXCL12-rich stroma surrounding the tumor impairs T cell tumor infiltration (Seo et al., 2019) by exerting a type of chemo-repulsion of CXCR4-expressing T cells (Vianello et al., 2006). Thus, targeting the CXCR4/CXCL12 axis can optimize intratumoral T cell localization, and promote T cell anti-tumor activity (Bockorny et al., 2020). Other chemokine receptors such as CCR5 (Spranger et al., 2015) and CCR6 (Bell et al., 1999) direct dendritic cell migration into tumors and improve the induction of anti-tumor immunity. In this regard, it has been shown that WNT/β-catenin pathway activation in some tumors correlates with immune T cell exclusion. This indicates that the development of pharmacologic inhibitors against this pathway could restore immune cell infiltration and improve the efficacy of immunotherapy (Luke et al., 2019).

Notably, the TME produces chemokines that attract immunosuppressive T cells. For example, CCR4-expressing regulatory T cells (T_{regs}) migrate very efficiently to tumors expressing the chemokines CCL17 and CCL22 (Curiel et al., 2004; Mizukami et al., 2008). A study demonstrated that inducing the expression of CCL22 at a site in the skin distant from the tumor in melanoma-challenged mice redirected T_{regs} away from the tumor and enhanced T cell anti-tumor responses (Klarquist et al., 2016). The anti-CCR4 antibody mogamulizumab was shown to be able to deplete T_{regs} in lung esophageal cancer patients (Kurose et al., 2015). In line with this, several CCR4 antagonists are now being tested in clinical trials although the therapeutic potential of CCR4 inhibition to deplete T_{regs} remains to be determined. Other chemokines upregulated in tumors such as CCL2 (Li et al., 2020) favor the recruitment of macrophages and myeloid-derived suppressor cells

---

Fig. 1 An overview of tumor microenvironment players affecting the efficacy of adoptive T cell therapies. Factors that can alter the expression of key elements to achieve transendothelial migration and the entry of lymphocytes into the tumor bed. The highly fibrotic structure of the stromal extracellular matrix (ECM) with the accumulation of type I collagen fibrils, hyaluronan (HA), fibronectin, laminins, fibulins or tenascins create a barrier for T cell infiltration. Aberrancies of endothelial vessels and adhesion molecule expression, immunosuppressive cytokines, chemokine-chemokine receptor mismatching, and recruitment of immunosuppressive cells such as cancer-associated fibroblast (CAF), Tumor associated macrophages (TAM), Myeloid derived suppressor cells (MDSC), tolerogenic dendritic cells (DC), neutrophils and T regulatory cells (Treg) greatly hinder the antitumor activity of T cells. Tumor cells alter their milieu by lowering oxygen, pH and glucose and increasing immunosuppressive metabolites, lactate and fatty acid levels that can inhibit the immune cells and facilitate tumor development.

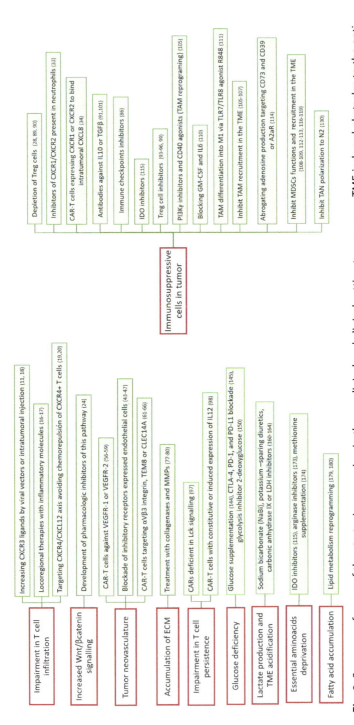

Fig. 2 Summary of some of the strategies emerging in the preclinical and clinical setting to overcome TME-induced obstacles to the anti-tumor activity of T cells.

expressing CCR2 and CCR5 (Jeong et al., 2019) while CXCL8 (Xie, 2001), guides the recruitment of CXCR1 and CXCR2-expressing neutrophils that may exhibit anti-tumor or pro-tumor features (Fridlender and Albelda, 2012). Several inhibitors of CXCR1/CXCR2 binding with cognate ligands are being developed (Dufies et al., 2019), although their use has still not been evaluated in the clinical setting. Whilding et al. have exploited the high levels of CXCL8 expressed in several tumors to improve the migratory capacity of CAR-T cells. Genetic modification of anti-tumor CAR-T cells to express CXCR1 or CXCR2 on their cell membrane improved their migratory capacity and enhanced their antitumor activity (Whilding et al., 2019). Targeting chemokines and receptors to improve cancer treatment is challenging, as expression is not exclusive to one immune cell type. Thus, the inhibition of a particular chemokine could have pleiotropic effects that require thorough consideration.

Other inflammatory mediators such as leukotrienes, thromboxanes and prostaglandins are also produced during tumor development. Their effects on vascular permeability and chemotaxis can favor the recruitment of immune cells to the tumor bed. In principle, inflammation is the body's protective response against various harmful stimuli. However, the aberrant and inappropriate chronic inflammation that occurs during tumor development becomes harmful and immunosuppressive. The presence of certain chemokines, cytokines, and myeloid cell subsets correlate with poor prognosis in colorectal cancers, as established in the "CRC immunoscore" (Mlecnik et al., 2016). Indeed, inflammation has been largely linked to tumor promotion since many inflammatory mediators can serve as direct growth factors for growing tumors, the stimulation of angiogenesis and recruitment of fibroblasts and other stromal cells, which exert tumor-supporting functions (Greten and Grivennikov, 2019). Through modulation of the TME, inflammatory mediators can alter the mechanical and metabolic properties of tumor stroma and cancer cells by altering the extracellular matrix, the availability of growth factors and key metabolites, as will be discussed in the following sections.

3. Physical barriers and the extracellular matrix

There are several factors that can alter the expression of key elements to achieve transendothelial migration and the entry of lymphocytes into the tumor bed (Lanitis et al., 2015). On the one hand, the down-regulation of the intercellular adhesion molecule 1 (ICAM-1) induced by vascular endothelial growth factor A (VEGF-A) and basic fibroblast growth factor (BFGF), and the overexpression and signaling via the endothelin-1/endothelin B-receptor axis help tumors evade T cell attack (Buckanovich et al., 2008; Griffioen et al., 1996; Tanaka et al., 2014). Indeed, overexpression of the endothelin B receptor has been associated with the absence of TILs and poor overall survival of the cancer patients. It has been shown that inhibitors of the endothelin B receptor can augment T cell adhesion to human endothelium in vitro and increase T cell homing to tumors in murine tumor models in a process requiring also ICAM-1 (Buckanovich et al., 2008). On the other hand, the upregulation of Fas ligand in response to VEGF-A, IL-10, and PGE2, specifically induces apoptosis of Fas-expressing $CD8^+$ T cells without affecting T_{regs} (Motz et al., 2014; Yu et al., 2003). Also, upregulation of a variety of inhibitory receptors such as B7-H3 (Zang et al., 2010), PD-L1 and PD-L2 (Rodig et al., 2003), Tim-3 (Huang et al., 2010; Wu et al., 2010) and B7-H4 (Krambeck et al., 2006) in endothelial cells, as well as the production of soluble inhibitory molecules including IL-6 (Neiva et al., 2014), PGE2 (Casos et al., 2011) and TGFβ (Casos et al., 2011; Pirtskhalaishvili and Nelson, 2000; Taflin et al., 2011), suppress T cell function and block their entry into the tumor bed (Joyce and Fearon, 2015).

Accelerated tumor expansion results in hypoxic conditions, leading to the activation of hypoxia-inducible factor-1α (HIF-1α), which in turn induces the upregulation of angiogenic factors such as VEGF, EGF, or FGF favoring aberrant hypervascularization (Bergers and Benjamin, 2003). The use of angiogenesis inhibitors to normalize tumor vasculature has attracted much attention. However, the activation of alternative angiogenic pathways by the tumor often results in resistance (Li et al., 2002). Immunological approaches to disrupting tumor vasculature have been successfully tested in pre-clinical studies (Li et al., 2002; van Beijnum et al., 2015). Notably, T cells expressing CARs targeting tumor vasculature antigens such as VEGFR-1 or VEGFR-2 demonstrated a significant delay in tumor growth in pre-clinical experiments (Chinnasamy et al., 2010; Hajari Taheri et al., 2019; Niederman et al., 2002; Wang et al., 2013b). However, tumor endothelial cells exhibit a remarkable heterogeneity, hence targeting one antigen might not be sufficient to achieve the desired antiangiogenic effect (Patten et al., 2010). Other studies used CAR-T cells to disrupt tumor neovasculature by targeting the αvβ3 integrin (Fu et al., 2013), TEM8 (Byrd et al., 2018; Petrovic et al., 2019), EDB alternative splicing of fibronectin (Wagner et al., 2021; Xie et al., 2019) or CLEC14A (Zhuang et al., 2020), although some toxicity issues arose (Chinnasamy et al., 2010; Petrovic et al., 2019) probably because of the "on-target off-tumor" activity of the CAR-T cells.

In addition to the aberrant vasculature, the fibrotic state of desmoplastic tumors can impair lymphocyte tumor infiltration. It is now well-accepted that carcinomas behave like wounds, which force the host TME into a constant state of fibrotic repair (Dvorak, 1986) with a continuous and extensive remodeling of extracellular matrix (ECM) due to the chronic inflammation present in the tumor. Tumor cells secrete growth factors to attract fibroblasts to the TME where they are transformed into cancer-associated fibroblasts (CAFs), the main actors of ECM accumulation. A particular feature of stromal ECM in many cancers is its highly fibrotic structure with the accumulation of type I collagen fibrils extensively crosslinked by lysyl oxidase (LOX) and tissue transglutaminase (TG2) (Perryman and Erler, 2014) that significantly affects tumor progression, metastasis and response-to-therapy (Jiang et al., 2017; Keely, 2011; Werb and Lu, 2015). These oriented collagen fibrils are a biomarker of poor outcome (Paszek et al., 2005). This collagen fibrillogenesis is supported by fibronectin (FN) networks that may provide a template for deposition of collagen and other components of the extracellular matrix including LTBP1, fibulin, tenascin, laminins, hyaluronan and thrombospondin among others, highlighting the interdependence of collagen and FN networks. Notably, FN in perivascular matrices constitutes an obligate scaffold for organization of the vessel-associated ECM and a repository for pro-angiogenic factors (Van Obberghen-Schilling et al., 2011). In immune-excluded tumors, there is an abundant infiltrate of immune cells that fail to effectively penetrate the tumor parenchyma and which remain in the stroma surrounding the tumor cell nests. This immune phenotype is characterized by an excessive deposition of ECM components, including dense aligned bundles of collagen and FN around tumor islets. Interestingly, it has been observed in live-cell imaging studies in tumor tissue sections that active T cell motility is enhanced in regions with less FN and collagen I, whereas T cells migrated poorly in dense matrix areas (Joyce and Fearon, 2015; Salmon et al., 2012). Also, the elevated accumulation of hyaluronan (HA) in the TME contributes to cancer progression and therapy resistance and it has been associated with lower T cell densities in the TME (Tahkola et al., 2021), supporting the idea that an HA-rich extracellular matrix can act as a shield between T cells and tumor cells preventing T cell infiltration into the TME. Thus, the high levels of collagen, FN and HA are responsible for the ECM stiffness that can promote tumor evasion from the immune system by limiting the contact of immune cells with cancer cells.

This excessive intratumoral deposition of collagen and HA can be attributed to accelerated synthesis and slowed catabolism. The enzymes catalyzing the degradation of the ECM tend to be suppressed in the TME (Hiltunen et al., 2002). For these reasons, ECM components have been considered as therapeutic targets for cancer. Tumor treatment with collagenases (Zinger et al., 2019) and MMPs (Leifler et al., 2013) to degrade collagen or hyaluronidases to reduce HA (Provenzano et al., 2012; Singha et al., 2015) were considered to improve tumor permeability or to improve the antitumor efficacy of CAR-T cells (Zhao et al., 2021). Migrating lymphocytes need cell-cell or cell-matrix interactions. Thus, the extracellular matrix may be seen as both a barrier and an important component to ensure forward propulsion. The repertoire of MMPs needed for the efficient migration of lymphocytes will likely vary with the ECM composition of the TME. The limited therapeutic efficacy seen in ACT might be due in part to the deficiency in the matrix degrading enzymes needed to degrade a specific TME (Owen et al., 2003; Zhao et al., 2021). However, cancer progression and metastasis are dependent on the proteolytic action of the MMPs present in the TME (Niland and Eble, 2020), so that their indiscriminate activation could favor the spread of the disease (Fang et al., 2014; Parks et al., 2004).

4. Tumor infiltrating immunosuppressor cells

The complexity of the extracellular matrix and the imbalance of chemokines in the TME greatly promote the presence of T_{regs}, tumor-associated macrophages (**TAMs**), myeloid suppressor cells (**MDSCs**) and other immunosuppressive populations, which greatly hinder the ability of T cells to recognize and remove tumor cells (Rodriguez-Garcia et al., 2020) affecting also to immune checkpoint therapies (Jia et al., 2020).

T regulatory cells: T_{regs} represent a crucial component of the immune system and are essential for controlling self-tolerance. However, they are capable of suppressing the function of tumor-reactive T cells. Indeed, T_{regs} are found infiltrating many types of human tumors and are associated with a high risk of death and reduced survival (reviewed in Lozano et al., 2013). Targeting mechanisms of T_{reg}-mediated immunosuppression is a reasonable approach to block/deplete T_{regs} and enhancing anti-tumor T cell responses. T_{regs} that infiltrate the tumors are characterized by the high expression of markers such as CTLA4, PD-1, LAG3, CD25, GITR, 41BB, ICOS, TIGIT, TIM3 and OX40 as compared to peripheral T_{regs} (Arce Vargas et al., 2018). These markers are potential targets to inhibit T_{regs} and work in synergy with ACT strategies (Arce Vargas et al., 2018; Chambers et al., 2001). Also, the use of antibodies against TGFβ to hamper T_{reg} recruitment has been shown to synergize with ACT therapies (Wallace et al., 2008). Tregs are characterized by the expression of CD25 and the Treg-specific FOXP3 transcription factor, which is required for their development and function (Williams and Rudensky, 2007). Several peptide inhibitors of FOXP3 have demonstrated the capacity to inhibit Treg cells in vivo improving their anti tumor activity in animal models (Casares et al., 2010; Lozano et al., 2015, 2017, 2021). Their potential synergy with ACT therapies remains to be elucidated.

Reasonably, the combination of ACT immunotherapy with T_{reg} inhibition strategies could improve the anti-tumor response. However, tumor preconditioning with lymphodepletion, by irradiation or cyclophosphamide, prior to ACT is still the most established recommendation. Nevertheless, CAR-T cell therapies remain vulnerable to multiple extrinsic immunosuppression factors including T_{regs}. CAR-T cells have been shown to inadvertently potentiate T_{regs} by providing a source of IL-2 for T_{reg} consumption. Some strategies such as third generation CARs which are deficient in Lck signaling are able to circumvent T_{regs} suppression (Suryadevara et al., 2019). Other approaches have focused on the enhancement of CAR-T cell resistance to suppression, rather than the direct targeting of suppressor cells. For example, a study using CAR-T cells with constitutive or induced expression of IL-12. IL-12 rendered transferred CAR-T cells less susceptible to T_{regs} (Pegram et al., 2012).

At present, multiple pharmaceutical and biotechnological companies are developing modern therapies to inhibit T_{regs} as bispecific antibodies, mimic antibodies and develop FOXP3 antisense oligonucleotide inhibitors, antibody combinations, and new conjugated drugs (Ohue and Nishikawa, 2019). However, only few of these have been tested in ACT immunotherapy models. In addition, immune homeostasis requires a balance between immune effector and suppressor mechanisms, and the systemic depletion of T_{regs} can result in the development of autoimmune adverse effects. Thus, selective depletion of tumor-infiltrating T_{regs} in the TME, while preserving the suppressive naïve and peripheral T_{reg} populations, will be the biggest challenge to increasing ACT antitumor activity without inducing detrimental autoimmunity (Romano et al., 2019).

Tumor-associated macrophages (TAMs). TAMs have been thought to be involved in tumor initiation, progression, angiogenesis and metastasis and are considered major players connecting inflammation and cancer. TAMs can be roughly classified into two contrasting groups: classically activated "M1" and alternatively activated "M2" macrophages. M2, and a small population of M1, exert anti-inflammatory and tumorigenic features. Drifting toward M2 may be a default pathway in macrophage differentiation into the TME. Among the factors able to influence macrophage polarization toward an M2 profile are IL-10, IL-6, TGF-β, and PGE2, all highly expressed in the TME (Allavena et al., 2008).

Numerous studies have shown the contribution of TAMs to immunotherapy resistance, although the precise mechanisms are still unclear. In the context of ACT, it is well described that TAMs contribute to building a more angiogenic and fibrotic TME by the production of factors such as VEGFA, PDGF, TGF-β and FGF. Moreover, the limited blood supply and abnormal structure of tumor neo-vessels restrict the traffic and infiltration of transferred cells, especially for TILs (Schaaf et al., 2018; Solinas et al., 2009).

Recent studies have partially addressed three major TAM-targeting strategies to boost ACT: macrophage elimination, recruitment inhibition, and reprogramming. Clinical trials have focused on the first two approaches because of the synergistic effect of macrophage-reprogramming agents, such as PI3Kγ inhibitors and CD40 agonists, need further analysis in animal models. Also, little is known about the intracellular metabolic TAMs switch during tumor progression and its potential impact on immunotherapy (Xiang et al., 2021). TAM elimination is usually based on bisphosphonates such as clodronate and zoledronate acid. Both strategies are effective in preclinical models but there is a collateral effect and tissue-resident macrophages might be also eliminated and overall homeostasis could be impaired as a result.

Recruitment of circulating monocytes is primarily dependent on chemokine signals and therefore the use of neutralizing monoclonal antibodies or small molecule inhibitors of these chemokines might be an effective way to prevent TAM accumulation in the TME (Tacke, 2017).

In this sense, CXCR4 and CCR2 inhibitors successfully impair TAM recruitment in animal models (Chen et al., 2015; Li et al., 2013).

Myeloid suppressor cells (MDSCs). Several studies have shown that MDSCs hinder ACT immunotherapy and inhibit both expansion and function of adoptively transferred T cells (Mengos et al., 2019). As is the case with TAMs, MDSCs can be targeted using different strategies: inhibiting their expansion, function and differentiation or depleting them. Inhibiting expansion and recruitment of MDSCs is achieved using chemokine inhibitors. CCR2 and CCR5 are essential in MDSC recruitment and notably, the CCR2 inhibitor *(CCX872) used in combination with anti PD-1 was found to* prolong mice survival in clinically relevant murine glioma models (Flores-Toro et al., 2020).

Other preclinical experiments showed that CAR-T cells increase MDSC levels in a GM-CSF- and IL-6 dependent manner, inhibiting the anti-tumor activity of adoptive T cells. Thus, blocking both these two cytokines is also a potentially useful procedure to consider (Burga et al., 2015).

TLR7/TLR8 activation *via* the TLR7/TLR8 agonist R848 induces M2 differentiation into anti-tumor M1-type macrophages. Thus, promoting differentiation of MDSC into macrophages and dendritic cells is also a promising future strategy (Liu et al., 2020). The MDSCs general functions are mediated by the STAT3 pathway. Likewise, strategies targeting STAT3 are under active investigation (Guha et al., 2019). At the intranuclear level, several histone deacetylase (HDAC) inhibitors are being studied to repress MDSC functions (Kim et al., 2020).

Regarding the inhibition of MDSC metabolism, on one hand, abrogating adenosine production by targeting the nucleotide-metabolizing enzymes CD73 and CD39 seems to be a promising therapeutic strategy (Li et al., 2018). On the other hand, indoleamine 2,3 dioxygenase (IDO), an intracellular enzyme in MDSCs that mediates tryptophan metabolism, hinders CAR-T therapy efficacy by the production of tryptophan metabolites. In a myeloma tumor model, it has been shown that IDO inhibitors could restore the therapeutic effect of transferred T cells (Xiang et al., 2020).

Removal of MDSCs has been also considered. Gemcitabine and 5-fluorouracil (5-FU) are the most common drugs used to eliminate MDSCs in tumor models and cancer patients (Wang et al., 2017). Also, high-dose radiation has been proposed as a candidate for MDSC elimination. Both chemotherapy and radiotherapy, in combination with tyrosine kinase inhibitors, PD-L1 blockade or the use of STING agonists, synergistically inhibited the infiltration of MDSCs, thus promoting antitumor immunity in ACT approaches (Deng et al., 2014; Hsu et al., 2015; Liang et al., 2017).

Of note, in murine tumors it has been shown that MDSCs can preserve their immunosuppressive properties without inhibiting the anti-tumor effect of memory, but not naive, transferred T cells. Thus, under certain

ACT conditions, myeloid cell immunosuppression could be overcome (Arina, 2014). In any case, it is well described that MDSCs are a cluster of heterogeneous cells and it is difficult to target them due to their diversity and their dynamic and plastic phenotype.

Cancer-associated fibroblasts (CAFs). CAFs are the most abundant stromal cell population within the TME in many solid tumors. In the last decade, CAFs have been highlighted as cells supporting cancer progression, metastasis and resistance to therapies through multidirectional signaling with tumor cells and other cells that play a role within the TME.

Moreover, CAFs have also been reported to alter immune cell functions by direct and indirect interactions. Specifically, CAFs can activate the expression of ECM proteins that act as a physical barrier to T cell infiltration. There are also certain chemokines secreted by CAFs such as CXCL12, which are able to sequester T cells surrounding tumors, preventing their recruitment and trafficking into the tumor site. In some human tumors, it has been reported that CAFs can directly upregulate exhaustion markers on T cells, such as PD-1, as well as interfere with T cell activation, inhibiting dendritic cell maturation or by direct inhibition of TCR signaling. Currently, CAF inhibition represents an unrecognized aspect on ACT and this area of research needs further attention in future studies (Freeman and Mielgo, 2020).

Mast cells (MCs). MCs have been associated with the regulation of Th2 immune responses by IgE secretion. But they also release a wide variety of molecules including proteases, lipids and many cytokines, chemokines and growth factors. Interestingly, diverse studies suggest a regulatory role for MCs via the regulation of T cell functions and also by impacting on regulatory T cells. Specifically, IL-6, TNFα, CCL5, leukotriene B4 and histamine can directly affect the recruitment, activation and proliferation of CD4 and CD8 T cells and T_{reg} (Bulfone-Paus and Bahri, 2015).

Although these consequences will be more relevant in inflammatory diseases, we cannot exclude the inflammatory component in solid tumors where hypoxia, adenosine, TLRs and complement, can activate MCs. Even if MCs have both pro-tumorigenic and anti-tumorigenic roles depending of TME composition or the cancer therapy employed, they are also candidates to consider in ACT strategies (Oldford and Marshall, 2015).

Tumor-associated neutrophils (TANs). TANs are emerging as one of the key tumor-infiltrating immune cells that influence cancer progression and metastasis. Recruited by cytokines, chemokines, lipids, and growth factors that are secreted from cancer cells and cancer-associated stromal cells,

TANs can exert immunosuppressive functions by releasing high levels of inducible nitric oxide synthase (iNOS), which inhibits proliferation of cytotoxic T cells (Coffelt et al., 2015; Lasarte et al., 1999). In fact, although TANs can also exert antitumor activities, the TME and TGFβ in particular, can favor neutrophil polarization from an N1 to N2 neutrophil phenotype with high pro-angiogenic, pro-metastatic and immunosuppressive activity (Shaul and Fridlender, 2018). The production of high levels of arginase, MMP-9 VEGF, and numerous chemokines by N2 neutrophils have been associated with poor clinical outcomes (Shaul and Fridlender, 2018). Neutrophils are important players supporting circulating tumor cells survival during their systemic dissemination and favoring the metastatic potential of circulating tumor cells (Szczerba et al., 2019). TANs release neutrophil extracellular traps (NETs) composed from nuclear or mitochondrial DNA decorated with proteases and inflammatory mediators, playing a key regulatory role of the tumor microenvironment (recently reviewed in Demkow, 2021; Teijeira et al., 2021). IL8 produced in the tumor recruit neutrophils to tumor lesions and triggers the extrusion of NETs. Thus, blockade of IL8 or its receptors (CXCR1 and CXCR2) is being pursued in clinical trials alone or in combination with anti-PD-L1 checkpoint inhibitors (Teijeira et al., 2021). Therapeutic strategies against TANs are based on the pharmacological blockade of tumor-derived factors that recruit and polarize neutrophils. Targeting the CXCL-8/CXCR-1/CXCR-2 axis (Gregory and Houghton, 2011) in TANs by modulating the TME substances affecting neutrophil plasticity has also been proposed.

5. Metabolic competition and interstitial fluid

Cancer cells with aberrant metabolism and excessive glucose uptake for aerobic glycolysis consume large amounts of oxygen and nutrients resulting in hypoxia, nutritional deficiency, and elevated level of metabolic by-products in the TME (Sahai et al., 2020). This highly immunosuppressive TME inhibits T cell function and induces immune escape. Several studies have shown that T cells infiltrating the tumor become anergic and exhausted, and present distinct metabolic signatures (Lim et al., 2020; Scharping et al., 2016; Siska et al., 2017). In this part of the review, we describe the effects of the different metabolic components of the TME on T cell function and anti-tumor immunity.

Hypoxia: The high metabolic rates of cancer cells in combination with poor vasculature make the TME highly hypoxic. Cancer cells can adapt to hypoxia and studies show that it could drive angiogenesis, metastasis, and chemoresistance (Harris, 2002; Wilson and Hay, 2011). The effect of hypoxia on tumor infiltrating T cell function and anti-tumor activity remains controversial. Under hypoxic conditions, HIF1α is released from its negative regulator von Hippel-Lindau (VHL) and upregulates the expression of glycolytic genes such as lactate dehydrogenase (LDH) and pyruvate dehydrogenase kinase (PDK1), and the glucose transporter GLUT-1. These genes increase glycolysis and reduce oxidative phosphorylation. HIF1α upregulates PD-L1 on MDSCs, which leads to T cell exhaustion, and can promote the generation of T_{regs} (Shen et al., 2019). On the other hand, hypoxic $CD8^+$ T cells are more cytotoxic against tumor cells by increasing Granzyme B packaging into granules (Gropper et al., 2017), and HIF1α promoted $CD8^+$ T cell activation and migration (Palazon et al., 2017). Liikanen et al. recently reported that VHL deficiency in $CD8^+$ CAR-T cells resulting in HIF activity enhanced accumulation of exhausted yet functional CAR-T cells in the tumor resulting in the higher efficacy of immunotherapy (Liikanen et al., 2021). HIF1α expression was reported to affect $CD4^+$ T cell differentiation. HIF1α targets FOXP3 for proteasomal degradation shifting the Treg:Th17 balance toward Th17 (Clambey et al., 2012). Also, HIF1α directly upregulates expression of IL-17 and RORγt (Dang et al., 2011). Overall, these findings indicate that hypoxia could promote T cell anti-tumor immunity and that there is a fine balance between.

Glucose deficiency: Upon TCR activation and CD28 co-stimulation, $CD4^+$ and $CD8^+$ T cells increase glycolysis and show enhanced expression of Glut-1 which makes them highly sensitive to glucose availability (Palmer et al., 2015). Due to both cancer cells and effector T cells undergoing aerobic glycolysis in the TME, there is strong competition for glucose, which impairs infiltrating T cell function and anti-tumor immunity. The high glycolytic rate of tumor cells limits the glucose available to T cells, which reduces the glycolytic intermediate phosphoenol-pyruvate limiting T cell receptor-mediated calcium signaling (Ho et al., 2015). Cham et al. performed a gene array on $CD8^+$ T cells under glucose deprivation conditions and observed that the expression of key cytokines such as IFNγ and Granzyme B and genes for cell cycle progression such as cyclin D2 were glucose-dependent (Cham et al., 2008). Lack of glucose also decreases mTOR activity reducing T cell function and differentiation and resulting in overall tumor progression (MacIver et al., 2013). The hypoglycemic TME has been shown to constrain the expression of the methyltransferase EZH2 in T cells (Zhao et al., 2016). EZH2 activates the Notch pathway and stimulates T cell polyfunctionality and survival via Bcl-2 (Zhao et al., 2016). EZH2 function and T cell polyfunctionality and survival were restored after glucose supplementation was provided for the T cells. It is worth mentioning that while providing T cells with supplemental glucose enhances their function, supplementing tumors with glucose increased tumorigenicity since it was the tumor cells that used up the glucose (Lin et al., 2020). Chang et al. showed that the glucose available to T cells in the TME can be restored by blocking CTLA-4, PD-1, and PD-L1 allowing T cell glycolysis and IFNγ production (Chang et al., 2015). Furthermore, blocking PD-L1 directly on tumors dampens glycolysis by reducing mTOR activity and enhances T cell anti-tumor immunity (Chang et al., 2015). Overall, these studies demonstrate the importance of glucose in mediating T cell function in the TME.

Modulating the metabolism of T cells is becoming an exciting avenue to improve current immunotherapies and improve T cell anti-tumor immunity in the adverse TME. Sukumar et al. treated T cells with the glycolysis inhibitor 2-deoxyglucose to enhance memory formation and further anti-tumor function (Sukumar et al., 2013). *Ex vivo* treatment of transferred T cells with an inhibitor of oxygen-sensing prolyl-hydroxylase domain proteins increased glycolytic activity and reduced melanoma lung metastasis in preclinical models (Clever et al., 2016). Preparing T cell metabolism *in vitro* prior to adoptive transfer is a way to enhance their anti-tumor activity *in vivo*.

Lactate and pH: The end products of glycolysis are lactate and protons (H^+), which are secreted massively by cancer cells and accumulate in the TME. Angelin et al. showed that lactate impairs effector T cell function and proliferation through LDH-mediated nicotinamide adenine dinucleotide (NAD^+) depletion (Angelin et al., 2017). LDH reduces NAD^+ to NADH and inhibits glycolysis in the NAD^+-dependent glyceraldehyde 3-phosphate dehydrogenase (GAPDH) reaction (Quinn et al., 2020). High lactate concentrations in the TME impede lactic acid export in $CD8^+$ T cells causing intracellular acidification, which suppresses NFAT induction, IFNγ expression, and anti-tumor immunity (Brand et al., 2016). Interestingly, $FOXP3^+$ T_{regs} which prefer oxidative phosphorylation over glycolysis are not negatively affected by high concentrations of lactate in the TME and have a metabolic advantage in low-glucose, lactate-rich environments (Angelin et al., 2017). In fact, a recent study demonstrated that T_{regs} upregulate pathways involved in the metabolism of lactate and that lactate is necessary for tumor infiltrating T_{reg} immunosuppressive function (Watson et al., 2021).

The accumulation of lactate and protons (H^+) in the TME causes acidification of the tumor. An intratumoral acidic pH of 6–6.5 is associated with metastasis, angiogenesis and therapy resistance, a characteristic phenotype of more aggressive tumors (Estrella et al., 2013; Garcia-Canaveras et al., 2019). Furthermore, acidic pH has detrimental effects on infiltrating T cell function. Acidic pH directly inhibits $CD8^+$ and $CD4^+$ T cell function and T cell mediated anti-tumor immunity resulting in immune escape and tolerance. Lowering the environmental pH to 6–6.5 induced anergy in $CD8^+$ T cells characterized by impairment of cytotoxic ability and reduced cytokine production and CD25 expression (Calcinotto et al., 2012). This anergic state was reversed by buffering pH to neutral values (Calcinotto et al., 2012). Exposing $CD8^+$ T cells to lactic acid caused a significant decrease in their proliferation, cytokine production, lytic activity, and tumor infiltration (Fischer et al., 2007). Furthermore, raising the TME pH by administration of sodium bicarbonate (NaBi) was shown to enhance anti-tumor immune responses in mouse models (Pilon-Thomas et al., 2016), and acute myeloid leukemia (AML) patients (Uhl et al., 2020). In AML, the lactic acid derived from the tumor caused extracellular acidosis, which subsequently lowered the intracellular pH of $CD8^+$ T cells reducing their proliferative and functional capacity (Uhl et al., 2020). This immunosuppressed T cell phenotype could be reversed by oral administration of NaBi, which also rescued the metabolic fitness of activated T cells and IFNγ production (Uhl et al., 2020). These data indicate that the negative effect the acidic TME has on infiltrating lymphocytes can be controlled and even reversed by basifying or alkalinizing the TME. There are other alternative pharmacologic approaches that could indirectly lead to the control of intracellular acidification. Potassium-sparing diuretics (amiloride) for metabolic alkalosis compensation (Poon et al., 2019), and the use of inhibitors for carbonic anhydrase IX (van Kuijk et al., 2016) or lactate dehydrogenase (LDH) (Oshima et al., 2020) are being considered to improve anti-tumor therapies. However, the efficacy of these agents to neutralize tumor pH or their effects on anti-tumor immunotherapy is not known.

Amino acids: In addition to the lack of available glucose, other essential nutrients for T cell function are limited in the TME. Sullivan et al. analyzed tumor interstitial fluid using mass spectrometry and observed that amino acids such as tryptophan (Trp) and arginine (Arg) were significantly reduced in the TME compared to plasma (Sullivan et al., 2019). Tumor cells express IDO, which mediates potent local effects on infiltrating T cells. IDO catalyzes the essential amino acid Trp via the kynurenine pathway. The rapid consumption of Trp from the TME mediates the activation of the serine/threonine-protein kinase GCN2 and inhibits the mTOR pathway (Metz et al., 2012). GCN2 activation causes cell-cycle arrest and anergy in $CD8^+$ T cells (Munn et al., 2005), and blocks $CD4^+$ Th17 differentiation promoting Treg differentiation and suppressive activity (Fallarino et al., 2006; Sundrud et al., 2009). Tumor cells, and specifically tumor stromal cells, express arginase, which is the enzyme that metabolizes Arg (Mussai et al., 2013). Reduced local Arg concentrations in the TME downregulates the expression of CD3ζ, which inhibits T cell proliferation and cytokine production upon TCR activation (Zabaleta et al., 2004). Lack of Arg also activates GCN2, which leads to cell cycle inhibitory effects on activated T cells similar to Trp depletion (Lee et al., 2003). Inhibition of arginase was able to modestly reduce the growth of lung carcinoma and enhance anti-tumor T cell responses. When it was combined with anti-PD-1 or STING agonist immunotherapy, it showed a significant improvement in survival outcomes (Sosnowska et al., 2021). A recent study has demonstrated the importance of methionine (Met) for T cell function in the TME. Tumor cells express high levels of the Met transporter SLC43A2 and outcompete $CD8^+$ T cells for Met. Disrupted Met metabolism in $CD8^+$ T cells results in the loss of H3K79me2, which leads to low STAT5 expression and inhibited T cell function (Bian et al., 2020). These studies show the importance of amino acids in the TME and how can we target their transport or metabolism to enhance anti-tumor T cell responses.

Fatty Acids: A hallmark of the TME is *de novo* lipogenesis due to the use of glucose or glutamine by tumor cells or adipose tumor tissue (Koundouros and Poulogiannis, 2020). This lipid synthesis results in the accumulation of fatty acids in the TME. Manzo et al. showed that in pancreatic tumors, there is an accumulation of long chain fatty acids creating a lipid-rich but nutrient-poor TME. Infiltrating $CD8^+$ T cells accumulate long chain fatty acids that rather than providing an energy source, mediate lipotoxicity and inhibit T cell anti-tumor function (Manzo et al., 2020). Furthermore, tumor infiltrating T_{regs} show enhanced lipid biosynthesis and rely on fatty acid oxidation more than conventional T_{regs}, indicating that the high concentration of fatty acid in the TME confers a preferential proliferative advantage to T_{regs} (Pacella et al., 2018). T_{regs} use extracellular fatty acids to suppress $CD8^+$ T cells (Miska et al., 2019). Thus, aberrant lipid metabolism is influencing anti-tumoral therapy response sustaining the emergence of resistance and consequently, lipid reprogramming strategies are being considered for therapeutic interventions in cancer (Bacci et al., 2021; Germain et al., 2020).

6. Concluding remarks

Data from preclinical studies and early clinical trials with adoptive T cell therapies have identified five major limitations in the context of solid tumors: T-cell recruitment, tumor cell recognition, activation and proliferation, persistence and bypass of the immunosuppressive TME. Most studies conducted to date have individually addressed these limitations as separate entities. However, the problem is so complex that it will be necessary to address different obstacles in parallel to increase the efficacy of T cells in solid tumors. An important approach will be the combination of several layers of engineering in the cellular product. Thanks to the rapid development of T-cell engineering and genome-editing technology, it is feasible to evaluate ambitious concepts, and to apply novel and often revolutionary clinical approaches, especially in the CAR-T cell field. However, some technical problems, such as the lower efficiency of transduction and transposition with increasing insert size, may limit the use of multi-armored CAR-T cells. On the other hand, the constitutive expression of therapeutics can further increase the toxicity of CAR-T cells and therefore it will be necessary to develop inducible expression strategies that allow the expression of the companion proteins in the TME. A promising scenario is the combination of CAR-T cells and other treatment modalities to maximize synergism and tackle different pathways simultaneously. Targeting the non-malignant cancer-associated stromal cells that promote tumor growth, metastasis and angiogenesis, or the extracellular matrix is another promising approach to remodeling the TME and facilitating the penetration of T cells into solid tumors. Driven mostly by the importance of demonstrating antitumor efficacy against human cancer cell lines, the clear majority of preclinical CAR-T cell validation experiments have been conducted in the context of immunodeficient mice tumor models. These models underestimate the effect of the immunosuppressive TME on adoptively transferred T cells. As a consequence, more appropriate preclinical models and mechanisms of efficacy and resistance to CAR-T cell therapy should be explored. Ongoing trials will reveal whether newer CAR-T cell approaches, including those that address the barriers to the TME, will benefit a broader population of cancer patients. The lessons we learn from these new clinical trials will be important in continuing to develop new CAR-T cell therapies for the treatment of solid tumors.

Acknowledgments

We are grateful to Dr. Paul Miller for English editing. The work was supported by grants from Ministerio de Ciencia e Innovación (PID2019-108989RB-I00, PLEC2021-008094 MCIN/AEI/10.13039/501100011033 and the European Union NextGenerationEU/ PRTR), Gobierno de Navarra (0011-1411-2019-000079; Proyecto DESCARTHeS) and Paula & Rodger Riney Foundation. The figures were created by the BioRender website platform.

References

Adams, D.H., Shaw, S., 1994. Leucocyte-endothelial interactions and regulation of leucocyte migration. Lancet 343 (8901), 831–836.

Allavena, P., Sica, A., Solinas, G., Porta, C., Mantovani, A., 2008. The inflammatory micro-environment in tumor progression: the role of tumor-associated macrophages. Crit. Rev. Oncol. Hematol. 66 (1), 1–9.

Angelin, A., Gil-de-Gomez, L., Dahiya, S., Jiao, J., Guo, L., Levine, M.H., et al., 2017. Foxp3 reprograms T cell metabolism to function in low-glucose, high-lactate environments. Cell Metab. 25 (6). 1282-93.e7.

Arce Vargas, F., Furness, A.J.S., Litchfield, K., Joshi, K., Rosenthal, R., Ghorani, E., et al., 2018. Fc effector function contributes to the activity of human anti-CTLA-4 antibodies. Cancer Cell 33 (4). 649-63.e4.

Arina, A., 2014. Rethinking the role of myeloid-derived suppressor cells in adoptive T-cell therapy for cancer. Oncoimmunology 3, e28464.

Bacci, M., Lorito, N., Smiriglia, A., Morandi, A., 2021. Fat and furious: lipid metabolism in antitumoral therapy response and resistance. Trends Cancer 7 (3), 198–213.

Bell, D., Chomarat, P., Broyles, D., Netto, G., Harb, G.M., Lebecque, S., et al., 1999. In breast carcinoma tissue, immature dendritic cells reside within the tumor, whereas mature dendritic cells are located in peritumoral areas. J. Exp. Med. 190 (10), 1417–1426.

Bellone, M., Calcinotto, A., 2013. Ways to enhance lymphocyte trafficking into tumors and fitness of tumor infiltrating lymphocytes. Front. Oncol. 3, 231.

Bergers, G., Benjamin, L.E., 2003. Tumorigenesis and the angiogenic switch. Nat. Rev. Cancer 3 (6), 401–410.

Bernhard, H., Neudorfer, J., Gebhard, K., Conrad, H., Hermann, C., Nahrig, J., et al., 2008. Adoptive transfer of autologous, HER2-specific, cytotoxic T lymphocytes for the treatment of HER2-overexpressing breast cancer. Cancer Immunol. Immunother. 57 (2), 271–280.

Bian, Y., Li, W., Kremer, D.M., Sajjakulnukit, P., Li, S., Crespo, J., et al., 2020. Cancer SLC43A2 alters T cell methionine metabolism and histone methylation. Nature 585 (7824), 277–282.

Bockorny, B., Semenisty, V., Macarulla, T., Borazanci, E., Wolpin, B.M., Stemmer, S.M., et al., 2020. BL-8040, a CXCR4 antagonist, in combination with pembrolizumab and chemotherapy for pancreatic cancer: the COMBAT trial. Nat. Med. 26 (6), 878–885.

Brand, A., Singer, K., Koehl, G.E., Kolitzus, M., Schoenhammer, G., Thiel, A., et al., 2016. LDHA-associated lactic acid production blunts tumor immunosurveillance by T and NK cells. Cell Metab. 24 (5), 657–671.

Buckanovich, R.J., Facciabene, A., Kim, S., Benencia, F., Sasaroli, D., Balint, K., et al., 2008. Endothelin B receptor mediates the endothelial barrier to T cell homing to tumors and disables immune therapy. Nat. Med. 14 (1), 28–36.

Bulfone-Paus, S., Bahri, R., 2015. Mast cells as regulators of T cell responses. Front. Immunol. 6, 394.

Burga, R.A., Thorn, M., Point, G.R., Guha, P., Nguyen, C.T., Licata, L.A., et al., 2015. Liver myeloid-derived suppressor cells expand in response to liver metastases in mice and inhibit the anti-tumor efficacy of anti-CEA CAR-T. Cancer Immunol. Immunother. 64 (7), 817–829.

Byrd, T.T., Fousek, K., Pignata, A., Szot, C., Samaha, H., Seaman, S., et al., 2018. TEM8/ANTXR1-specific CAR T cells as a targeted therapy for triple-negative breast cancer. Cancer Res. 78 (2), 489–500.

Calcinotto, A., Filipazzi, P., Grioni, M., Iero, M., De Milito, A., Ricupito, A., et al., 2012. Modulation of microenvironment acidity reverses anergy in human and murine tumor-infiltrating T lymphocytes. Cancer Res. 72 (11), 2746–2756.

Casares, N., Rudilla, F., Arribillaga, L., Llopiz, D., Riezu-Boj, J.I., Lozano, T., et al., 2010. A peptide inhibitor of FOXP3 impairs regulatory T cell activity and improves vaccine efficacy in mice. J. Immunol. 185 (9), 5150–5159.

Casos, K., Siguero, L., Fernandez-Figueras, M.T., Leon, X., Sarda, M.P., Vila, L., et al., 2011. Tumor cells induce COX-2 and mPGES-1 expression in microvascular endothelial cells mainly by means of IL-1 receptor activation. Microvasc. Res. 81 (3), 261–268.

Castermans, K., Griffioen, A.W., 2007. Tumor blood vessels, a difficult hurdle for infiltrating leukocytes. Biochim. Biophys. Acta 1776 (2), 160–174.

Cham, C.M., Driessens, G., O'Keefe, J.P., Gajewski, T.F., 2008. Glucose deprivation inhibits multiple key gene expression events and effector functions in CD8+ T cells. Eur. J. Immunol. 38 (9), 2438–2450.

Chambers, C.A., Kuhns, M.S., Egen, J.G., Allison, J.P., 2001. CTLA-4-mediated inhibition in regulation of T cell responses: mechanisms and manipulation in tumor immunotherapy. Annu. Rev. Immunol. 19, 565–594.

Chang, C.H., Qiu, J., O'Sullivan, D., Buck, M.D., Noguchi, T., Curtis, J.D., et al., 2015. Metabolic competition in the tumor microenvironment is a driver of cancer progression. Cell 162 (6), 1229–1241.

Chen, Y., Ramjiawan, R.R., Reiberger, T., Ng, M.R., Hato, T., Huang, Y., et al., 2015. CXCR4 inhibition in tumor microenvironment facilitates anti-programmed death receptor-1 immunotherapy in sorafenib-treated hepatocellular carcinoma in mice. Hepatology 61 (5), 1591–1602.

Chheda, Z.S., Sharma, R.K., Jala, V.R., Luster, A.D., Haribabu, B., 2016. Chemoattractant receptors BLT1 and CXCR3 regulate antitumor immunity by facilitating CD8+ T cell migration into tumors. J. Immunol. 197 (5), 2016–2026.

Chinnasamy, D., Yu, Z., Theoret, M.R., Zhao, Y., Shrimali, R.K., Morgan, R.A., et al., 2010. Gene therapy using genetically modified lymphocytes targeting VEGFR-2 inhibits the growth of vascularized syngenic tumors in mice. J. Clin. Invest. 120 (11), 3953–3968.

Clambey, E.T., McNamee, E.N., Westrich, J.A., Glover, L.E., Campbell, E.L., Jedlicka, P., et al., 2012. Hypoxia-inducible factor-1 alpha-dependent induction of FoxP3 drives regulatory T-cell abundance and function during inflammatory hypoxia of the mucosa. Proc. Natl. Acad. Sci. U. S. A. 109 (41), E2784–E2793.

Clever, D., Roychoudhuri, R., Constantinides, M.G., Askenase, M.H., Sukumar, M., Klebanoff, C.A., et al., 2016. Oxygen sensing by T cells establishes an immunologically tolerant metastatic niche. Cell 166 (5), 1117-31.e14.

Coffelt, S.B., Kersten, K., Doornebal, C.W., Weiden, J., Vrijland, K., Hau, C.S., et al., 2015. IL-17-producing gammadelta T cells and neutrophils conspire to promote breast cancer metastasis. Nature 522 (7556), 345–348.

Curiel, T.J., Coukos, G., Zou, L., Alvarez, X., Cheng, P., Mottram, P., et al., 2004. Specific recruitment of regulatory T cells in ovarian carcinoma fosters immune privilege and predicts reduced survival. Nat. Med. 10 (9), 942–949.

Dang, E.V., Barbi, J., Yang, H.Y., Jinasena, D., Yu, H., Zheng, Y., et al., 2011. Control of T(H)17/T(reg) balance by hypoxia-inducible factor 1. Cell 146 (5), 772–784.

Demkow, U., 2021. Neutrophil extracellular traps (NETs) in cancer invasion, evasion and metastasis. Cancers (Basel) 13 (17), 4495.

Deng, L., Liang, H., Burnette, B., Beckett, M., Darga, T., Weichselbaum, R.R., et al., 2014. Irradiation and anti-PD-L1 treatment synergistically promote antitumor immunity in mice. J. Clin. Invest. 124 (2), 687–695.

Dufies, M., Grytsai, O., Ronco, C., Camara, O., Ambrosetti, D., Hagege, A., et al., 2019. New CXCR1/CXCR2 inhibitors represent an effective treatment for kidney or head and neck cancers sensitive or refractory to reference treatments. Theranostics 9 (18), 5332–5346.

Dvorak, H.F., 1986. Tumors: wounds that do not heal. Similarities between tumor stroma generation and wound healing. N. Engl. J. Med. 315 (26), 1650–1659.

Estrella, V., Chen, T., Lloyd, M., Wojtkowiak, J., Cornnell, H.H., Ibrahim-Hashim, A., et al., 2013. Acidity generated by the tumor microenvironment drives local invasion. Cancer Res. 73 (5), 1524–1535.

Fallarino, F., Grohmann, U., You, S., McGrath, B.C., Cavener, D.R., Vacca, C., et al., 2006. The combined effects of tryptophan starvation and tryptophan catabolites down-regulate T cell receptor zeta-chain and induce a regulatory phenotype in naive T cells. J. Immunol. 176 (11), 6752–6761.

Fang, M., Yuan, J., Peng, C., Li, Y., 2014. Collagen as a double-edged sword in tumor progression. Tumour Biol. 35 (4), 2871–2882.

Fischer, K., Hoffmann, P., Voelkl, S., Meidenbauer, N., Ammer, J., Edinger, M., et al., 2007. Inhibitory effect of tumor cell-derived lactic acid on human T cells. Blood 109 (9), 3812–3819.

Flores-Toro, J.A., Luo, D., Gopinath, A., Sarkisian, M.R., Campbell, J.J., Charo, I.F., et al., 2020. CCR2 inhibition reduces tumor myeloid cells and unmasks a checkpoint inhibitor effect to slow progression of resistant murine gliomas. Proc. Natl. Acad. Sci. U. S. A. 117 (2), 1129–1138.

Freeman, P., Mielgo, A., 2020. Cancer-associated fibroblast mediated inhibition of CD8+ cytotoxic t cell accumulation in tumours: mechanisms and therapeutic opportunities. Cancers (Basel) 12 (9).

Fridlender, Z.G., Albelda, S.M., 2012. Tumor-associated neutrophils: friend or foe? Carcinogenesis 33 (5), 949–955.

Fu, X., Rivera, A., Tao, L., Zhang, X., 2013. Genetically modified T cells targeting neovasculature efficiently destroy tumor blood vessels, shrink established solid tumors and increase nanoparticle delivery. Int. J. Cancer 133 (10), 2483–2492.

Garcia-Canaveras, J.C., Chen, L., Rabinowitz, J.D., 2019. The tumor metabolic microenvironment: lessons from lactate. Cancer Res. 79 (13), 3155–3162.

Germain, N., Dhayer, M., Boileau, M., Fovez, Q., Kluza, J., Marchetti, P., 2020. Lipid metabolism and resistance to anticancer treatment. Biology (Basel) 9 (12), 474.

Gregory, A.D., Houghton, A.M., 2011. Tumor-associated neutrophils: new targets for cancer therapy. Cancer Res. 71 (7), 2411–2416.

Greten, F.R., Grivennikov, S.I., 2019. Inflammation and cancer: triggers, mechanisms, and consequences. Immunity 51 (1), 27–41.

Griffioen, A.W., Damen, C.A., Martinotti, S., Blijham, G.H., Groenewegen, G., 1996. Endothelial intercellular adhesion molecule-1 expression is suppressed in human malignancies: the role of angiogenic factors. Cancer Res. 56 (5), 1111–1117.

Gropper, Y., Feferman, T., Shalit, T., Salame, T.M., Porat, Z., Shakhar, G., 2017. Culturing CTLs under hypoxic conditions enhances their cytolysis and improves their anti-tumor function. Cell Rep. 20 (11), 2547–2555.

Guha, P., Gardell, J., Darpolor, J., Cunetta, M., Lima, M., Miller, G., et al., 2019. STAT3 inhibition induces Bax-dependent apoptosis in liver tumor myeloid-derived suppressor cells. Oncogene 38 (4), 533–548.

Hajari Taheri, F., Hassani, M., Sharifzadeh, Z., Behdani, M., Arashkia, A., Abolhassani, M., 2019. T cell engineered with a novel nanobody-based chimeric antigen receptor against VEGFR2 as a candidate for tumor immunotherapy. IUBMB Life 71 (9), 1259–1267.

Harris, A.L., 2002. Hypoxia—a key regulatory factor in tumour growth. Nat. Rev. Cancer 2 (1), 38–47.

Hensbergen, P.J., Wijnands, P.G., Schreurs, M.W., Scheper, R.J., Willemze, R., Tensen, C.P., 2005. The CXCR3 targeting chemokine CXCL11 has potent antitumor activity in vivo involving attraction of CD8+ T lymphocytes but not inhibition of angiogenesis. J. Immunother. 28 (4), 343–351.

Hiltunen, E.L., Anttila, M., Kultti, A., Ropponen, K., Penttinen, J., Yliskoski, M., et al., 2002. Elevated hyaluronan concentration without hyaluronidase activation in malignant epithelial ovarian tumors. Cancer Res. 62 (22), 6410–6413.

Ho, P.C., Bihuniak, J.D., Macintyre, A.N., Staron, M., Liu, X., Amezquita, R., et al., 2015. Phosphoenolpyruvate is a metabolic checkpoint of anti-tumor T cell responses. Cell 162 (6), 1217–1228.

Hogg, N., Patzak, I., Willenbrock, F., 2011. The insider's guide to leukocyte integrin signalling and function. Nat. Rev. Immunol. 11 (6), 416–426.

Hong, M., Puaux, A.L., Huang, C., Loumagne, L., Tow, C., Mackay, C., et al., 2011. Chemotherapy induces intratumoral expression of chemokines in cutaneous melanoma, favoring T-cell infiltration and tumor control. Cancer Res. 71 (22), 6997–7009.

Hsu, F.T., Chen, T.C., Chuang, H.Y., Chang, Y.F., Hwang, J.J., 2015. Enhancement of adoptive T cell transfer with single low dose pretreatment of doxorubicin or paclitaxel in mice. Oncotarget 6 (42), 44134–44150.

Huang, X., Bai, X., Cao, Y., Wu, J., Huang, M., Tang, D., et al., 2010. Lymphoma endothelium preferentially expresses Tim-3 and facilitates the progression of lymphoma by mediating immune evasion. J. Exp. Med. 207 (3), 505–520.

Jeong, J., Suh, Y., Jung, K., 2019. Context drives diversification of monocytes and neutrophils in orchestrating the tumor microenvironment. Front. Immunol. 10, 1817.

Jia, Y., Liu, L., Shan, B., 2020. Future of immune checkpoint inhibitors: focus on tumor immune microenvironment. Ann. Transl. Med. 8 (17), 1095.

Jiang, H., Hegde, S., DeNardo, D.G., 2017. Tumor-associated fibrosis as a regulator of tumor immunity and response to immunotherapy. Cancer Immunol. Immunother. 66 (8), 1037–1048.

John, L.B., Devaud, C., Duong, C.P., Yong, C.S., Beavis, P.A., Haynes, N.M., et al., 2013. Anti-PD-1 antibody therapy potently enhances the eradication of established tumors by gene-modified T cells. Clin. Cancer Res. 19 (20), 5636–5646.

Joyce, J.A., Fearon, D.T., 2015. T cell exclusion, immune privilege, and the tumor microenvironment. Science 348 (6230), 74–80.

Keely, P.J., 2011. Mechanisms by which the extracellular matrix and integrin signaling act to regulate the switch between tumor suppression and tumor promotion. J. Mammary Gland Biol. Neoplasia 16 (3), 205–219.

Kim, Y.D., Park, S.M., Ha, H.C., Lee, A.R., Won, H., Cha, H., et al., 2020. HDAC inhibitor, CG-745, enhances the anti-cancer effect of anti-PD-1 immune checkpoint inhibitor by modulation of the immune microenvironment. J. Cancer 11 (14), 4059–4072.

Kistner, L., Doll, D., Holtorf, A., Nitsche, U., Janssen, K.P., 2017. Interferon-inducible CXC-chemokines are crucial immune modulators and survival predictors in colorectal cancer. Oncotarget 8 (52), 89998–90012.

Klarquist, J., Tobin, K., Farhangi Oskuei, P., Henning, S.W., Fernandez, M.F., Dellacecca, E.R., et al., 2016. Ccl22 diverts T regulatory cells and controls the growth of melanoma. Cancer Res. 76 (21), 6230–6240.

Kohli, K., Pillarisetty, V.G., Kim, T.S., 2022. Key chemokines direct migration of immune cells in solid tumors. Cancer Gene Ther. 29 (1), 10–21.

Koundouros, N., Poulogiannis, G., 2020. Reprogramming of fatty acid metabolism in cancer. Br. J. Cancer 122 (1), 4–22.

Krambeck, A.E., Thompson, R.H., Dong, H., Lohse, C.M., Park, E.S., Kuntz, S.M., et al., 2006. B7-H4 expression in renal cell carcinoma and tumor vasculature: associations with cancer progression and survival. Proc. Natl. Acad. Sci. U. S. A. 103 (27), 10391–10396.

Kurose, K., Ohue, Y., Wada, H., Iida, S., Ishida, T., Kojima, T., et al., 2015. Phase Ia study of FoxP3+ CD4 Treg depletion by infusion of a humanized anti-CCR4 antibody, KW-0761, in cancer patients. Clin. Cancer Res. 21 (19), 4327–4336.

Lanitis, E., Irving, M., Coukos, G., 2015. Targeting the tumor vasculature to enhance T cell activity. Curr. Opin. Immunol. 33, 55–63.

Lasarte, J.J., Corrales, F.J., Casares, N., Lopez-Diaz de Cerio, A., Qian, C., Xie, X., et al., 1999. Different doses of adenoviral vector expressing IL-12 enhance or depress the immune response to a coadministered antigen: the role of nitric oxide. J. Immunol. 162 (9), 5270–5277.

Lawrence, M.B., Kansas, G.S., Kunkel, E.J., Ley, K., 1997. Threshold levels of fluid shear promote leukocyte adhesion through selectins (CD62L,P,E). J. Cell Biol. 136 (3), 717–727.

Lee, J., Ryu, H., Ferrante, R.J., Morris Jr., S.M., Ratan, R.R., 2003. Translational control of inducible nitric oxide synthase expression by arginine can explain the arginine paradox. Proc. Natl. Acad. Sci. U. S. A. 100 (8), 4843–4848.

Leifler, K.S., Svensson, S., Abrahamsson, A., Bendrik, C., Robertson, J., Gauldie, J., et al., 2013. Inflammation induced by MMP-9 enhances tumor regression of experimental breast cancer. J. Immunol. 190 (8), 4420–4430.

Li, Y., Wang, M.N., Li, H., King, K.D., Bassi, R., Sun, H., et al., 2002. Active immunization against the vascular endothelial growth factor receptor flk1 inhibits tumor angiogenesis and metastasis. J. Exp. Med. 195 (12), 1575–1584.

Li, M., Knight, D.A., L AS, Smyth MJ, Stewart TJ., 2013. A role for CCL2 in both tumor progression and immunosurveillance. Oncoimmunology 2 (7), e25474.

Li, L., Wang, L., Li, J., Fan, Z., Yang, L., Zhang, Z., et al., 2018. Metformin-induced reduction of CD39 and CD73 blocks myeloid-derived suppressor cell activity in patients with ovarian cancer. Cancer Res. 78 (7), 1779–1791.

Li, B.H., Garstka, M.A., Li, Z.F., 2020. Chemokines and their receptors promoting the recruitment of myeloid-derived suppressor cells into the tumor. Mol. Immunol. 117, 201–215.

Liang, H., Deng, L., Hou, Y., Meng, X., Huang, X., Rao, E., et al., 2017. Host STING-dependent MDSC mobilization drives extrinsic radiation resistance. Nat. Commun. 8 (1), 1736.

Liikanen, I., Lauhan, C., Quon, S., Omilusik, K., Phan, A.T., Bartroli, L.B., Ferry, A., Goulding, J., Chen, J., Scott-Browne, J.P., Yustein, J.T., Scharping, N.E., Witherden, D.A., Goldrath, A.W., 2021. Hypoxia-inducible factor activity promotes antitumor effector function and tissue residency by CD8+ T cells. J. Clin. Invest. 131 (7), e143729.

Lim, A.R., Rathmell, W.K., Rathmell, J.C., 2020. The tumor microenvironment as a metabolic barrier to effector T cells and immunotherapy. Elife 9, e55185. https://doi.org/10.7554/eLife.55185.

Lin, X., Xiao, Z., Chen, T., Liang, S.H., Guo, H., 2020. Glucose metabolism on tumor plasticity, diagnosis, and treatment. Front. Oncol. 10, 317.

Liu, Z., Xie, Y., Xiong, Y., Liu, S., Qiu, C., Zhu, Z., et al., 2020. TLR 7/8 agonist reverses oxaliplatin resistance in colorectal cancer via directing the myeloid-derived suppressor cells to tumoricidal M1-macrophages. Cancer Lett. 469, 173–185.

Lozano, T., Casares, N., Lasarte, J.J., 2013. Searching for the Achilles heel of FOXP3. Front. Oncol. 3, 294.

Lozano, T., Villanueva, L., Durantez, M., Gorraiz, M., Ruiz, M., Belsue, V., et al., 2015. Inhibition of FOXP3/NFAT interaction enhances T cell function after TCR stimulation. J. Immunol. 195 (7), 3180–3189.

Lozano, T., Gorraiz, M., Lasarte-Cia, A., Ruiz, M., Rabal, O., Oyarzabal, J., Hervás-Stubbs, S., Llopiz, D., Sarobe, P., Prieto, J., Casares, N., Lasarte, J.J., 2017. Blockage of FOXP3 transcription factor dimerization and FOXP3/AML1 interaction inhibits T regulatory cell activity: sequence optimization of a peptide inhibitor. Oncotarget 8 (42), 71709–71724. https://doi.org/10.18632/oncotarget.17845.

Lozano, T., Casares, N., Martil-Otal, C., Anega, B., Gorraiz, M., Parker, J., et al., 2021. Searching for peptide inhibitors of T regulatory cell activity by targeting specific domains of FOXP3 transcription factor. Biomedicine 9 (2).

Luke, J.J., Bao, R., Sweis, R.F., Spranger, S., Gajewski, T.F., 2019. WNT/beta-catenin pathway activation correlates with immune exclusion across human cancers. Clin. Cancer Res. 25 (10), 3074–3083.

MacIver, N.J., Michalek, R.D., Rathmell, J.C., 2013. Metabolic regulation of T lymphocytes. Annu. Rev. Immunol. 31, 259–283.

Manzo, T., Prentice, B.M., Anderson, K.G., Raman, A., Schalck, A., Codreanu, G.S., Nava Lauson, C.B., Tiberti, S., Raimondi, A., Jones, M.A., Reyzer, M., Bates, B.M., Spraggins, J.M., Patterson, N.H., McLean, J.A., Rai, K., Tacchetti, C., Tucci, S., Wargo, J.A., Rodighiero, S., Clise-Dwyer, K., Sherrod, S.D., Kim, M., Navin, N.E., Caprioli, R.M., Greenberg, P.D., Draetta, G., Nezi, L., 2020. Accumulation of long-chain fatty acids in the tumor microenvironment drives dysfunction in intrapancreatic CD8+ T cells. J. Exp. Med. 217 (8), e20191920. https://doi.org/10.1084/jem.20191920.

Mauldin, I.S., Wages, N.A., Stowman, A.M., Wang, E., Smolkin, M.E., Olson, W.C., et al., 2016. Intratumoral interferon-gamma increases chemokine production but fails to increase T cell infiltration of human melanoma metastases. Cancer Immunol. Immunother. 65 (10), 1189–1199.

Mengos, A.E., Gastineau, D.A., Gustafson, M.P., 2019. The CD14(+)HLA-DR(lo/neg) monocyte: an immunosuppressive phenotype that restrains responses to cancer immunotherapy. Front. Immunol. 10, 1147.

Metz, R., Rust, S., Duhadaway, J.B., Mautino, M.R., Munn, D.H., Vahanian, N.N., et al., 2012. IDO inhibits a tryptophan sufficiency signal that stimulates mTOR: a novel IDO effector pathway targeted by D-1-methyl-tryptophan. Oncoimmunology 1 (9), 1460–1468.

Mikucki, M.E., Fisher, D.T., Matsuzaki, J., Skitzki, J.J., Gaulin, N.B., Muhitch, J.B., et al., 2015. Non-redundant requirement for CXCR3 signalling during tumoricidal T-cell trafficking across tumour vascular checkpoints. Nat. Commun. 6, 7458.

Miska, J., Lee-Chang, C., Rashidi, A., Muroski, M.E., Chang, A.L., Lopez-Rosas, A., et al., 2019. HIF-1alpha is a metabolic switch between glycolytic-driven migration and oxidative phosphorylation-driven immunosuppression of Tregs in glioblastoma. Cell Rep. 27 (1). 226-37.e4.

Mizukami, Y., Kono, K., Kawaguchi, Y., Akaike, H., Kamimura, K., Sugai, H., et al., 2008. CCL17 and CCL22 chemokines within tumor microenvironment are related to accumulation of Foxp3+ regulatory T cells in gastric cancer. Int. J. Cancer 122 (10), 2286–2293.

Mlecnik, B., Bindea, G., Angell, H.K., Maby, P., Angelova, M., Tougeron, D., et al., 2016. Integrative analyses of colorectal cancer show immunoscore is a stronger predictor of patient survival than microsatellite instability. Immunity 44 (3), 698–711.

Motz, G.T., Santoro, S.P., Wang, L.P., Garrabrant, T., Lastra, R.R., Hagemann, I.S., et al., 2014. Tumor endothelium FasL establishes a selective immune barrier promoting tolerance in tumors. Nat. Med. 20 (6), 607–615.

Munn, D.H., Sharma, M.D., Baban, B., Harding, H.P., Zhang, Y., Ron, D., et al., 2005. GCN2 kinase in T cells mediates proliferative arrest and anergy induction in response to indoleamine 2,3-dioxygenase. Immunity 22 (5), 633–642.

Mussai, F., De Santo, C., Abu-Dayyeh, I., Booth, S., Quek, L., McEwen-Smith, R.M., et al., 2013. Acute myeloid leukemia creates an arginase-dependent immunosuppressive microenvironment. Blood 122 (5), 749–758.

Neiva, K.G., Warner, K.A., Campos, M.S., Zhang, Z., Moren, J., Danciu, T.E., et al., 2014. Endothelial cell-derived interleukin-6 regulates tumor growth. BMC Cancer 14, 99.

Niederman, T.M., Ghogawala, Z., Carter, B.S., Tompkins, H.S., Russell, M.M., Mulligan, R.C., 2002. Antitumor activity of cytotoxic T lymphocytes engineered to target vascular endothelial growth factor receptors. Proc. Natl. Acad. Sci. U. S. A. 99 (10), 7009–7014.

Niland, S., Eble, J.A., 2020. Hold on or cut? Integrin- and MMP-mediated cell-matrix interactions in the tumor microenvironment. Int. J. Mol. Sci. 22 (1), 238. https://doi.org/10.3390/ijms22010238.

Ohue, Y., Nishikawa, H., 2019. Regulatory T (Treg) cells in cancer: can Treg cells be a new therapeutic target? Cancer Sci. 110 (7), 2080–2089.

Oldford, S.A., Marshall, J.S., 2015. Mast cells as targets for immunotherapy of solid tumors. Mol. Immunol. 63 (1), 113–124.

Oshima, N., Ishida, R., Kishimoto, S., Beebe, K., Brender, J.R., Yamamoto, K., et al., 2020. Dynamic imaging of LDH inhibition in tumors reveals rapid in vivo metabolic rewiring and vulnerability to combination therapy. Cell Rep. 30 (6), 1798–810 e4.

Owen, J.L., Iragavarapu-Charyulu, V., Gunja-Smith, Z., Herbert, L.M., Grosso, J.F., Lopez, D.M., 2003. Up-regulation of matrix metalloproteinase-9 in T lymphocytes of mammary tumor bearers: role of vascular endothelial growth factor. J. Immunol. 171 (8), 4340–4351.

Pacella, I., Procaccini, C., Focaccetti, C., Miacci, S., Timperi, E., Faicchia, D., et al., 2018. Fatty acid metabolism complements glycolysis in the selective regulatory T cell expansion during tumor growth. Proc. Natl. Acad. Sci. U. S. A. 115 (28), E6546–E6555.

Palazon, A., Tyrakis, P.A., Macias, D., Velica, P., Rundqvist, H., Fitzpatrick, S., et al., 2017. An HIF-1alpha/VEGF-A axis in cytotoxic T cells regulates tumor progression. Cancer Cell 32 (5). 669-83.e5.

Palmer, C.S., Ostrowski, M., Balderson, B., Christian, N., Crowe, S.M., 2015. Glucose metabolism regulates T cell activation, differentiation, and functions. Front. Immunol. 6, 1.

Parks, W.C., Wilson, C.L., Lopez-Boado, Y.S., 2004. Matrix metalloproteinases as modulators of inflammation and innate immunity. Nat. Rev. Immunol. 4 (8), 617–629.

Paszek, M.J., Zahir, N., Johnson, K.R., Lakins, J.N., Rozenberg, G.I., Gefen, A., et al., 2005. Tensional homeostasis and the malignant phenotype. Cancer Cell 8 (3), 241–254.

Patten, S.G., Adamcic, U., Lacombe, K., Minhas, K., Skowronski, K., Coomber, B.L., 2010. VEGFR2 heterogeneity and response to anti-angiogenic low dose metronomic cyclophosphamide treatment. BMC Cancer 10, 683.

Pegram, H.J., Lee, J.C., Hayman, E.G., Imperato, G.H., Tedder, T.F., Sadelain, M., et al., 2012. Tumor-targeted T cells modified to secrete IL-12 eradicate systemic tumors without need for prior conditioning. Blood 119 (18), 4133–4141.

Perryman, L., Erler, J.T., 2014. Lysyl oxidase in cancer research. Future Oncol. 10 (9), 1709–1717.

Petrovic, K., Robinson, J., Whitworth, K., Jinks, E., Shaaban, A., Lee, S.P., 2019. TEM8/ANTXR1-specific CAR T cells mediate toxicity in vivo. PLoS One 14 (10), e0224015.

Pilon-Thomas, S., Kodumudi, K.N., El-Kenawi, A.E., Russell, S., Weber, A.M., Luddy, K., et al., 2016. Neutralization of tumor acidity improves antitumor responses to immunotherapy. Cancer Res. 76 (6), 1381–1390.

Pirtskhalaishvili, G., Nelson, J.B., 2000. Endothelium-derived factors as paracrine mediators of prostate cancer progression. Prostate 44 (1), 77–87.

Pockaj, B.A., Sherry, R.M., Wei, J.P., Yannelli, J.R., Carter, C.S., Leitman, S.F., et al., 1994. Localization of 111indium-labeled tumor infiltrating lymphocytes to tumor in patients receiving adoptive immunotherapy. Augmentation with cyclophosphamide and correlation with response. Cancer 73 (6), 1731–1737.

Poon, A.C., Inkol, J.M., Luu, A.K., Mutsaers, A.J., 2019. Effects of the potassium-sparing diuretic amiloride on chemotherapy response in canine osteosarcoma cells. J. Vet. Intern. Med. 33 (2), 800–811.

Provenzano, P.P., Cuevas, C., Chang, A.E., Goel, V.K., Von Hoff, D.D., Hingorani, S.R., 2012. Enzymatic targeting of the stroma ablates physical barriers to treatment of pancreatic ductal adenocarcinoma. Cancer Cell 21 (3), 418–429.

Quinn 3rd, W.J., Jiao, J., TeSlaa, T., Stadanlick, J., Wang, Z., Wang, L., et al., 2020. Lactate limits T cell proliferation via the NAD(H) redox state. Cell Rep. 33 (11), 108500.

Rodig, N., Ryan, T., Allen, J.A., Pang, H., Grabie, N., Chernova, T., et al., 2003. Endothelial expression of PD-L1 and PD-L2 down-regulates CD8+ T cell activation and cytolysis. Eur. J. Immunol. 33 (11), 3117–3126.

Rodriguez-Garcia, A., Palazon, A., Noguera-Ortega, E., Powell Jr., D.J., Guedan, S., 2020. CAR-T cells hit the tumor microenvironment: strategies to overcome tumor escape. Front. Immunol. 11, 1109.

Romano, M., Fanelli, G., Albany, C.J., Giganti, G., Lombardi, G., 2019. Past, present, and future of regulatory T cell therapy in transplantation and autoimmunity. Front. Immunol. 10, 43.

Sahai, E., Astsaturov, I., Cukierman, E., DeNardo, D.G., Egeblad, M., Evans, R.M., et al., 2020. A framework for advancing our understanding of cancer-associated fibroblasts. Nat. Rev. Cancer 20 (3), 174–186.

Salmon, H., Franciszkiewicz, K., Damotte, D., Dieu-Nosjean, M.C., Validire, P., Trautmann, A., et al., 2012. Matrix architecture defines the preferential localization and migration of T cells into the stroma of human lung tumors. J. Clin. Invest. 122 (3), 899–910.

Schaaf, M.B., Garg, A.D., Agostinis, P., 2018. Defining the role of the tumor vasculature in antitumor immunity and immunotherapy. Cell Death Dis. 9 (2), 115.

Scharping, N.E., Menk, A.V., Moreci, R.S., Whetstone, R.D., Dadey, R.E., Watkins, S.C., et al., 2016. The tumor microenvironment represses T cell mitochondrial biogenesis to drive intratumoral T cell metabolic insufficiency and dysfunction. Immunity 45 (2), 374–388.

Seo, Y.D., Jiang, X., Sullivan, K.M., Jalikis, F.G., Smythe, K.S., Abbasi, A., et al., 2019. Mobilization of CD8(+) T cells via CXCR4 blockade facilitates PD-1 checkpoint therapy in human pancreatic cancer. Clin. Cancer Res. 25 (13), 3934–3945.

Shaul, M.E., Fridlender, Z.G., 2018. Cancer-related circulating and tumor-associated neutrophils—subtypes, sources and function. FEBS J. 285 (23), 4316–4342.

Shen, X., Zhang, L., Li, J., Li, Y., Wang, Y., Xu, Z.X., 2019. Recent findings in the regulation of programmed death ligand 1 expression. Front. Immunol. 10, 1337.

Singha, N.C., Nekoroski, T., Zhao, C., Symons, R., Jiang, P., Frost, G.I., et al., 2015. Tumor-associated hyaluronan limits efficacy of monoclonal antibody therapy. Mol. Cancer Ther. 14 (2), 523–532.

Siska, P.J., Beckermann, K.E., Mason, F.M., Andrejeva, G., Greenplate, A.R., Sendor, A.B., Chiang, Y.J., Corona, A.L., Gemta, L.F., Vincent, B.G., Wang, R.C., Kim, B., Hong, J., Chen, C.L., Bullock, T.N., Irish, J.M., Rathmell, W.K., Rathmell, J.C., 2017. Mitochondrial dysregulation and glycolytic insufficiency functionally impair CD8 T cells infiltrating human renal cell carcinoma. JCI Insight 2 (12), e93411. https://doi.org/10.1172/jci.insight.93411.

Solinas, G., Germano, G., Mantovani, A., Allavena, P., 2009. Tumor-associated macrophages (TAM) as major players of the cancer-related inflammation. J. Leukoc. Biol. 86 (5), 1065–1073.

Sosnowska, A., Chlebowska-Tuz, J., Matryba, P., Pilch, Z., Greig, A., Wolny, A., et al., 2021. Inhibition of arginase modulates T-cell response in the tumor microenvironment of lung carcinoma. Oncoimmunology 10 (1), 1956143.

Spranger, S., Bao, R., Gajewski, T.F., 2015. Melanoma-intrinsic beta-catenin signalling prevents anti-tumour immunity. Nature 523 (7559), 231–235.

Sukumar, M., Liu, J., Ji, Y., Subramanian, M., Crompton, J.G., Yu, Z., et al., 2013. Inhibiting glycolytic metabolism enhances CD8+ T cell memory and antitumor function. J. Clin. Invest. 123 (10), 4479–4488.

Sullivan, M.R., Danai, L.V., Lewis, C.A., Chan, S.H., Gui, D.Y., Kunchok, T., Dennstedt, E.A., Vander Heiden, M.G., Muir, A., 2019. Quantification of microenvironmental metabolites in murine cancers reveals determinants of tumor nutrient availability. Elife 8, e44235. https://doi.org/10.7554/eLife.44235.

Sundrud, M.S., Koralov, S.B., Feuerer, M., Calado, D.P., Kozhaya, A.E., Rhule-Smith, A., et al., 2009. Halofuginone inhibits TH17 cell differentiation by activating the amino acid starvation response. Science 324 (5932), 1334–1338.

Suryadevara, C.M., Desai, R., Farber, S.H., Choi, B.D., Swartz, A.M., Shen, S.H., et al., 2019. Preventing Lck activation in CAR T cells confers Treg resistance but requires 4-1BB signaling for them to persist and treat solid tumors in nonlymphodepleted hosts. Clin. Cancer Res. 25 (1), 358–368.

Szczerba, B.M., Castro-Giner, F., Vetter, M., Krol, I., Gkountela, S., Landin, J., et al., 2019. Neutrophils escort circulating tumour cells to enable cell cycle progression. Nature 566 (7745), 553–557.

Tacke, F., 2017. Targeting hepatic macrophages to treat liver diseases. J. Hepatol. 66 (6), 1300–1312.

Taflin, C., Favier, B., Baudhuin, J., Savenay, A., Hemon, P., Bensussan, A., et al., 2011. Human endothelial cells generate Th17 and regulatory T cells under inflammatory conditions. Proc. Natl. Acad. Sci. U. S. A. 108 (7), 2891–2896.

Tahkola, K., Ahtiainen, M., Mecklin, J.P., Kellokumpu, I., Laukkarinen, J., Tammi, M., et al., 2021. Stromal hyaluronan accumulation is associated with low immune response and poor prognosis in pancreatic cancer. Sci. Rep. 11 (1), 12216.

Tanaka, T., Sho, M., Takayama, T., Wakatsuki, K., Matsumoto, S., Migita, K., et al., 2014. Endothelin B receptor expression correlates with tumour angiogenesis and prognosis in oesophageal squamous cell carcinoma. Br. J. Cancer 110 (4), 1027–1033.

Teijeira, A., Garasa, S., Ochoa, M.C., Villalba, M., Olivera, I., Cirella, A., et al., 2021. IL8, neutrophils, and NETs in a collusion against cancer immunity and immunotherapy. Clin. Cancer Res. 27 (9), 2383–2393.

Uhl, F.M., Chen, S., O'Sullivan, D., Edwards-Hicks, J., Richter, G., Haring, E., Andrieux, G., Halbach, S., Apostolova, P., Büscher, J., Duquesne, S., Melchinger, W., Sauer, B., Shoumariyeh, K., Schmitt-Graeff, A., Kreutz, M., Lübbert, M., Duyster, J., Brummer, T., Boerries, M., Madl, T., Blazar, B.R., Groß, O., Pearce, E.L., Zeiser, R., 2020. Metabolic reprogramming of donor T cells enhances graft-versus-leukemia effects in mice and humans. Sci. Transl. Med. 12 (567), eabb8969. https://doi.org/10.1126/scitranslmed.abb8969.

van Beijnum, J.R., Nowak-Sliwinska, P., Huijbers, E.J., Thijssen, V.L., Griffioen, A.W., 2015. The great escape; the hallmarks of resistance to antiangiogenic therapy. Pharmacol. Rev. 67 (2), 441–461.

van Kuijk, S.J., Gieling, R.G., Niemans, R., Lieuwes, N.G., Biemans, R., Telfer, B.A., et al., 2016. The sulfamate small molecule CAIX inhibitor S4 modulates doxorubicin efficacy. PLoS One 11 (8), e0161040.

Van Obberghen-Schilling, E., Tucker, R.P., Saupe, F., Gasser, I., Cseh, B., Orend, G., 2011. Fibronectin and tenascin-C: accomplices in vascular morphogenesis during development and tumor growth. Int. J. Dev. Biol. 55 (4–5), 511–525.

Vaupel, P., Hockel, M., 2000. Blood supply, oxygenation status and metabolic micromilieu of breast cancers: characterization and therapeutic relevance. Int. J. Oncol. 17 (5), 869–879.

Vianello, F., Papeta, N., Chen, T., Kraft, P., White, N., Hart, W.K., et al., 2006. Murine B16 melanomas expressing high levels of the chemokine stromal-derived factor-1/CXCL12 induce tumor-specific T cell chemorepulsion and escape from immune control. J. Immunol. 176 (5), 2902–2914.

Wagner, J., Wickman, E., Shaw, T.I., Anido, A.A., Langfitt, D., Zhang, J., et al., 2021. Antitumor effects of CAR T cells redirected to the EDB splice variant of fibronectin. Cancer Immunol. Res. 9 (3), 279–290.

Wallace, A., Kapoor, V., Sun, J., Mrass, P., Weninger, W., Heitjan, D.F., et al., 2008. Transforming growth factor-beta receptor blockade augments the effectiveness of adoptive T-cell therapy of established solid cancers. Clin. Cancer Res. 14 (12), 3966–3974.

Wang, X., Lu, X.L., Zhao, H.Y., Zhang, F.C., Jiang, X.B., 2013a. A novel recombinant protein of IP10-EGFRvIIIscFv and CD8(+) cytotoxic T lymphocytes synergistically inhibits the growth of implanted glioma in mice. Cancer Immunol. Immunother. 62 (7), 1261–1272.

Wang, W., Ma, Y., Li, J., Shi, H.S., Wang, L.Q., Guo, F.C., et al., 2013b. Specificity redirection by CAR with human VEGFR-1 affinity endows T lymphocytes with tumor-killing ability and anti-angiogenic potency. Gene Ther. 20 (10), 970–978.

Wang, Z., Till, B., Gao, Q., 2017. Chemotherapeutic agent-mediated elimination of myeloid-derived suppressor cells. Oncoimmunology 6 (7), e1331807.

Watson, M.J., Vignali, P.D.A., Mullett, S.J., Overacre-Delgoffe, A.E., Peralta, R.M., Grebinoski, S., et al., 2021. Metabolic support of tumour-infiltrating regulatory T cells by lactic acid. Nature 591 (7851), 645–651.

Werb, Z., Lu, P., 2015. The role of stroma in tumor development. Cancer J. 21 (4), 250–253.

Whilding, L.M., Halim, L., Draper, B., Parente-Pereira, A.C., Zabinski, T., Davies, D.M., Maher, J., 2019. CAR T-cells targeting the integrin αvβ6 and co-expressing the chemokine receptor CXCR2 demonstrate enhanced homing and efficacy against several solid malignancies. Cancers (Basel) 11 (5), 674. https://doi.org/10.3390/cancers11050674.

Williams, L.M., Rudensky, A.Y., 2007. Maintenance of the Foxp3-dependent developmental program in mature regulatory T cells requires continued expression of Foxp3. Nat. Immunol. 8 (3), 277–284.

Wilson, W.R., Hay, M.P., 2011. Targeting hypoxia in cancer therapy. Nat. Rev. Cancer 11 (6), 393–410.

Wu, F.H., Yuan, Y., Li, D., Lei, Z., Song, C.W., Liu, Y.Y., et al., 2010. Endothelial cell-expressed Tim-3 facilitates metastasis of melanoma cells by activating the NF-kappaB pathway. Oncol. Rep. 24 (3), 693–699.

Xiang, X., He, Q., Ou, Y., Wang, W., Wu, Y., 2020. Efficacy and safety of CAR-modified T cell therapy in patients with relapsed or refractory multiple myeloma: a meta-analysis of prospective clinical trials. Front. Pharmacol. 11, 544754.

Xiang, X., Wang, J., Lu, D., Xu, X., 2021. Targeting tumor-associated macrophages to synergize tumor immunotherapy. Signal Transduct. Target. Ther. 6 (1), 75.

Xie, K., 2001. Interleukin-8 and human cancer biology. Cytokine Growth Factor Rev. 12 (4), 375–391.

Xie, Y.J., Dougan, M., Jailkhani, N., Ingram, J., Fang, T., Kummer, L., et al., 2019. Nanobody-based CAR T cells that target the tumor microenvironment inhibit the growth of solid tumors in immunocompetent mice. Proc. Natl. Acad. Sci. U. S. A. 116 (16), 7624–7631.

Yu, J.S., Lee, P.K., Ehtesham, M., Samoto, K., Black, K.L., Wheeler, C.J., 2003. Intratumoral T cell subset ratios and Fas ligand expression on brain tumor endothelium. J. Neurooncol. 64 (1–2), 55–61.

Zabaleta, J., McGee, D.J., Zea, A.H., Hernandez, C.P., Rodriguez, P.C., Sierra, R.A., et al., 2004. Helicobacter pylori arginase inhibits T cell proliferation and reduces the expression of the TCR zeta-chain (CD3zeta). J. Immunol. 173 (1), 586–593.

Zang, X., Sullivan, P.S., Soslow, R.A., Waitz, R., Reuter, V.E., Wilton, A., et al., 2010. Tumor associated endothelial expression of B7-H3 predicts survival in ovarian carcinomas. Mod. Pathol. 23 (8), 1104–1112.

Zhang, S., Kohli, K., Black, R.G., Yao, L., Spadinger, S.M., He, Q., et al., 2019. Systemic interferon-gamma increases MHC class I expression and T-cell infiltration in cold tumors: results of a phase 0 clinical trial. Cancer Immunol. Res. 7 (8), 1237–1243.

Zhao, E., Maj, T., Kryczek, I., Li, W., Wu, K., Zhao, L., et al., 2016. Cancer mediates effector T cell dysfunction by targeting microRNAs and EZH2 via glycolysis restriction. Nat. Immunol. 17 (1), 95–103.

Zhao, R., Cui, Y., Zheng, Y., Li, S., Lv, J., Wu, Q., et al., 2021. Human hyaluronidase PH20 potentiates the antitumor activities of mesothelin-specific CAR-T cells against gastric cancer. Front. Immunol. 12, 660488.

Zhuang, X., Maione, F., Robinson, J., Bentley, M., Kaul, B., Whitworth, K., Jumbu, N., Jinks, E., Bystrom, J., Gabriele, P., Garibaldi, E., Delmastro, E., Nagy, Z., Gilham, D., Giraudo, E., Bicknell, R., Lee, S.P., 2020. CAR T cells targeting tumor endothelial marker CLEC14A inhibit tumor growth. JCI Insight 5 (19), e138808. https://doi.org/10.1172/jci.insight.138808.

Zinger, A., Koren, L., Adir, O., Poley, M., Alyan, M., Yaari, Z., et al., 2019. Collagenase nanoparticles enhance the penetration of drugs into pancreatic tumors. ACS Nano 13 (10), 11008–11021.

CHAPTER TWO

Dendritic cell transfer for cancer immunotherapy

Liwei Zhao[a,b], Shuai Zhang[a,b], Oliver Kepp[a,b,*], Guido Kroemer[a,b,c,*], and Peng Liu[a,b,*]

[a]Metabolomics and Cell Biology Platforms, Gustave Roussy Cancer Center, Université Paris Saclay, Villejuif, France
[b]Centre de Recherche des Cordeliers, Equipe labellisée par la Ligue contre le cancer, Université de Paris, Sorbonne Université, Inserm U1138, Institut Universitaire de France, Paris, France
[c]Institut du Cancer Paris Carpem, Department of Biology, Hôpital Européen Georges Pompidou, APHP, Paris, France
*Corresponding authors: e-mail address: captain.olsen@gmail.com; kroemer@orange.fr; kunzhiling@gmail.com

Contents

1. Introduction 34
2. Morphological and functional characteristics of DC 35
3. Role of DCs in cancer immunotherapy 36
    3.1 DC-mediated phagocytosis 37
    3.2 Priming of lymphocytes by DCs 38
    3.3 DC as tumoricidal effector cells 39
4. Concept of DC-based immunotherapies 40
    4.1 Generation of autologous DCs for adaptive transfer 41
    4.2 Preclinical model cell lines for cancer-relevant DC studies 44
5. Experimental and clinical approaches to improve DC-based immunotherapies 46
    5.1 Combination with adjuvants to potentiate DC-mediated immune responses 47
    5.2 Exclusion of immunosuppressive factors to restore DC-mediated immune responses 49
    5.3 In cooperation with other immunotherapies 50
6. Engineering dendritic cells to improve DC-based cancer immunotherapy 51
7. Concluding remarks 52
Acknowledgments 54
Conflict of interest 54
References 54

Abstract

Dendritic cells (DCs) play a major role in cancer immunosurveillance as they bridge innate and adaptive immunity by detecting tumor-associated antigens and presenting them to T lymphocytes. The adoptive transfer of antigen loaded DCs has been proposed as an immunotherapeutic approach for the treatment of various types of cancer.

Nevertheless, despite promising preclinical data, the therapeutic efficacy of DC transfer is still deceptive in cancer patients. Here we summarize recent findings in DC biology with a special focus on the development of actionable therapeutic strategies and discuss experimental and clinical approaches that aim at improving the efficacy of DC-based immunotherapies, including, but not limited to, optimized DC production and antigen loading, stimulated maturation, the co-treatment with additional immunotherapies, as well as the inhibition of DC checkpoints.

1. Introduction

Continuously occurring genetic mutations in proliferative tissues can give rise to nascent tumors. Malignant transformation is generally accompanied by the generation of tumor-associated antigens (TAAs) which can be detected by the innate immune system ultimately leading to the eradication of arising neoplasms (Sanmamed and Chen, 2018). However, under certain conditions, immunosurveillance mechanisms fail, which inevitably leads to the establishment of cancer. In most cases, immunologically silent transformation involves the suppression of the host immune system through a variety of mechanisms, which—together with an acquired resistance against immune attack—allows tumors to thrive.

Tumor immune escape mechanisms include, but are not limited to, the downregulation of surface antigens reducing tumor immunogenicity, the upregulation of immune checkpoints inhibiting T cell activation, the recruitment of immunosuppressive cells and the production of immune-toxic metabolites dampening the tumoricidal immune response (Pham et al., 2018).

Many immune escape mechanisms involve actionable targets for the development of immunotherapeutic strategies such as antitumor vaccination, immune checkpoint blockade (ICB), chimeric antigen receptor T cells (CAR-T), as well as the use of therapeutic cytokines. Most of these strategies aim at improving T functionality, yet mandatorily depend on the functional presentation of TAAs by professional antigen-presenting cells (APCs).

APCs facilitate the capture, processing and presentation of TAAs via major histocompatibility complex (MHC) class II molecules together with costimulatory molecules to T cells (de Jong et al., 2006; Kambayashi and Laufer, 2014). B cells, macrophages and dendritic cells (DCs) can act as APCs, yet only activated DCs have the ability to migrate to draining lymph nodes and present the captured TAA to naïve T cells. In addition, the capacity of DCs to present antigen surpasses that of macrophages and B cells by

100 to 1,000 times (Levin et al., 1993; Paglia et al., 1996), making them a key player for the initiation of antigen-specific immunity, and the most potent candidates for viable approaches to enforce antigen-specific and therapeutically relevant T cell responses. In this review, we will discuss the biological and immunological characteristics of DCs and the therapeutic potential of targeting DCs for cancer immunotherapy.

2. Morphological and functional characteristics of DC

DCs were first described by Ralph M. Steinman in peripheral lymphoid organs from mice, and were named for their stellate pleomorphism or branch-like (dendritic) protrusions, which become visible in the mature state (Steinman and Cohn, 1973). Following the morphological description, a series of studies by Steinman and others revealed the role of DCs in initiating T cell-mediated immune responses. As a bridge between innate and adaptive immunity, DCs serve as elite sentinels that migrate between lymphoid tissues, capture, process and present antigens, thereby activating specific T cell responses (Steinman and Hemmi, 2006).

DCs are a heterogenous population of cells sparsely distributed across various lymphoid and non-lymphoid tissues, that share common features. However, DCs differ in phenotype, surface markers, migration pattern, cytokine production and immune function at different developmental stages (Sichien et al., 2017). Human DCs can be divided into myeloid DC and lymphoid DC, both of which originate from bone marrow-resident $CD34^+$ hematopoietic stem cells (Banchereau et al., 2000; Diao et al., 2006; Naik et al., 2006). Myeloid DCs are habitually referred to as conventional DCs (cDCs), which can be further subdivided into type 1 cDCs (cDC1) and type 2 cDCs (cDC2), based on their surface markers and transcriptomic signatures (Collin and Bigley, 2018; Merad et al., 2013) Lymphoid DCs, also called plasmacytoid DCs (pDC), were originally characterized as lymphoid cells negative for lineage markers of B, T or nature killer (NK) cells (Grouard et al., 1997). pDCs become potent producers of type I interferons upon stimulation of toll like receptors (TLRs) during viral infection, and are therefore critical for antiviral immunity (Koucky et al., 2019).

Immature DCs can be recruited to peripheral tissues in response to immunogenic stimuli such as pathogen associated molecular patterns (PAMPs) arising from microbial infection or danger associated molecular patterns (DAMPs) released from dying cells. Such PAMPs and DAMPs are recognized by pattern recognition receptors (PRRs) expressed on DCs. DAMP-mediated PRR signaling at sites of inflammation includes, but is not limited to, the ligand-receptor pairs ATP-P2RX7, ANXA1-FPR1, CALR-CD91 and HMGB1-TLR4 (Apetoh et al., 2007; Ghiringhelli et al., 2009; Obeid et al., 2007; Vacchelli et al., 2015). Furthermore, CCR7, which is expressed upon stimulation of pathogenic or inflammatory signals, can interact with CCL19 and CCL21, which guide DC via afferent lymphatics towards the lymph nodes (Liu et al., 2021). Immature DCs have a large phagocytic potential and the ability to engulf antigen yet are relatively poor antigen presenters. The engagement with, and internalization of antigen, in parallel to PRR signaling triggers DC maturation and their migration to secondary lymphoid tissues. Mature DCs lose most of their phagocytic capability while acquiring strong antigen presenting properties, thus evolving into immunologically competent DCs (Thery and Amigorena, 2001).

During maturation DCs undergo a complex series of phenotypic changes resulting in functional alterations. Thus, while immature DCs express only low levels of MHC molecules, costimulatory molecules (such as CD80 and CD86), chemokine receptors (including CCR7), and adhesion molecules (such as CD40 and CD44,) mature DCs upregulate the expression of these surface molecules and acquire additional markers including CD11c and CD83, that are essential for their migration to the lymphoid tissues and optimal activation of cytotoxic T cell response (Sheng et al., 2005). In addition, mature DCs secrete cytokines including interleukin 1 beta (IL-1β), IL-6, IL-12, tumor necrosis factor-α (TNF-α) and type I interferons (IFN), to control the activation or polarization of other lymphocytes, including T helper cells (Th1/Th2/Th17) (Maldonado-Lopez and Moser, 2001; Terhune et al., 2013), regulatory T cells (Tregs) (Ghiringhelli et al., 2005), B cells (Jego et al., 2005), NK cells (Munz et al., 2005) and NKT cells (Fujii et al., 2002). This implies the multidirectional involvement of DCs in the regulation of both innate and adaptive immune responses.

3. Role of DCs in cancer immunotherapy

Their central role in immunosurveillance provides a convincing rational for employing DCs as a tool for eliciting endogenous anti-tumor responses for effective tumor eradication. Indeed, numerous studies revealed that the level of DC infiltration into solid tumors negatively correlates with disease progression in many types of cancer (Tran Janco et al., 2015), and positively correlates with long-term prognosis and survival of cancer patients

(Aarntzen et al., 2012; Ishigami et al., 2000; Roberts et al., 2016; Treilleux et al., 2004). The general principle of DC-based immunotherapy is based on the *ex vivo* differentiation of autologous DCs precursors together with their maturation/activation by the loading with specific TAAs or the exposure to whole tumor lysate. Mature DCs can then be re-transferred to the organism to generate an immune response aiming at the elimination of cancer cells.

The therapeutic potential of transferred DCs can unfold by mobilizing multiple arms of the antitumor immune response, including, but not restricted to, the direct engulfing of tumor cells or inhibition of tumor cell growth via direct engagement, the priming of lymphocytes and/or the secretion of cytotoxic or regulatory cytokines, chemokines, and DC-derived exosomes (Chan and Housseau, 2008; van Beek et al., 2014; Xu et al., 2020).

3.1 DC-mediated phagocytosis

DCs, especially in their immature state, are important members of the group of professional phagocytes and are able to accomplish cellular engulfment at high efficiency. Phagocytosis is defined as the uptake of relatively large particles ($>0.5\,\mu m$), including microorganisms and apoptotic cells (Rabinovitch, 1995). Although not as potent as macrophages, immature DCs can efficiently ingest dying (in most cases apoptotic or necroptotic) cancer cells. It has been well described that cancer cells that die from immunogenic cell death (ICD), for instance in response to cytotoxicants or ionizing radiation, expose a series of DAMPs that are recognized by corresponding PRRs, thereby triggering cellular engulfment by DCs (Apetoh et al., 2007; Ghiringhelli et al., 2009; Kroemer et al., 2013; Obeid et al., 2007; Vacchelli et al., 2015). The most prominent DAMPs exposed by dying cancer cells include calreticulin (CALR) (Obeid et al., 2007) and heat shock proteins (HSP) (Basu et al., 2001), that can both ligate CD91, as well as phosphatidylserine (PS), which binds to PS receptors (Park and Kim, 2017). In experimental settings, engineered viable tumor cells that constitutively expose those DAMPs on their surface are efficiently phagocytosed by DCs (Chan et al., 2007; Xu et al., 2017). Engulfed dying cancer cells or parts thereof are embedded in the phagosome, which then undergoes a series of maturation steps before fusing with lysosomes to form phagolysosomes, in which the ingested cells are eventually digested (Savina and Amigorena, 2007). Of note, different subpopulations of DCs vary in surface expression of phagocytic receptors, which favor the selective uptake of microorganisms and particles. For example, $CD8^+$ cDC1 are more potent in phagocytosing apoptotic cells than other

DC subtypes present in the same location (Iyoda et al., 2002), which is in support of their central role in cross-presenting tumor antigens to cytotoxic T lymphocytes (CTLs) thus igniting antitumor immunity.

3.2 Priming of lymphocytes by DCs

It is important to emphasize that only immature DCs are referred to as phagocytic, and that immature DCs undergo programmed maturation once they are engaged with pathogens or cancer cells that are emitting DAMPs, ultimately resulting in the irreversible loss of their phagocytic capacity (Savina and Amigorena, 2007). As DCs mature they acquire an exquisite capacity of antigen presentation and hence become professional APCs. Dying cancer cells and soluble tumor antigens from living cancer cells are major sources of TAAs for processing and presentation by DCs (Regnault et al., 1999; Sallusto et al., 1995). Apoptotic cell bodies captured through phagocytosis and soluble tumor antigens acquired via Fc- or C-type lectin receptor-mediated macropinocytosis or endocytosis are processed into antigenic peptide fragments, and then loaded on MHC molecules. Once such peptide-MHC complexes reach their cell surface, DCs can prime specific $CD4^+$ and $CD8^+$ T cells for the initiation of adaptive immune responses (Albert et al., 1998; Geijtenbeek et al., 2004; Guermonprez et al., 2002; Platt et al., 2010). Indeed, very few presented peptides from complex proteins are sufficient for T cell recognition.

For $CD4^+$ T cell immunity, large antigen fragments are loaded on MHC-II molecules and are further processed into shorter peptides of 10–20 amino acid length. The short peptide-MHC II complexes translocate to the cell surface and are presented to T cell receptors (TCRs) on $CD4^+$ T cells to promote activation, proliferation, and differentiation. Typically, cytokines produced by mature DCs as well as by activated macrophages, neutrophils and B cells, together with upregulated costimulatory molecules allow for the priming of naïve $CD4^+$ T cells and direct their polarization into different subtypes of helper cells (Th) such as Th1, Th2, Th17 that subsequently promote other immune cells to amplify antitumor immunity (Heufler et al., 1996; Tay et al., 2021).

Tumor infiltrating $CD8^+$ cytotoxic T cells are the main executioners of cancer cells in the context of antitumor immunity. Tumor antigen-presenting DCs migrate to the draining lymph nodes, where the DC costimulatory molecules engage with TCRs expressed by naïve $CD8^+$ T cells to drive their differentiation into effector T cells, then proceed to robust activation and proliferation. Of note, this antigen-specific $CD8^+$ T cell activation can also happen directly in the tumor bed, likely within tertiary lymphoid structures involving the participation of tumor infiltrating DCs (Bai et al., 2001; Wolkers et al., 2001). Apart from this classical cross-presentation of antigens to cytotoxic T cells, some special exosomes carrying abundant MHC and costimulatory molecules can derive from tumor antigen-sensitized DCs and migrate to the tumor bed or lymphoid organs with processed antigen for indirect or direct induction of T cell immunity (Rao et al., 2016; Tkach et al., 2017). Effector $CD8^+$ T cells are typically regarded as a group of cytotoxic cells that produce IFN γ, TNF-α, and granzyme B, which together execute acute tumoricidal tasks (Farhood et al., 2019). In addition to the proliferation and activation of cytotoxic T cells for acute tumor cell killing, the cross-presentation of tumor antigens by DCs also facilitates the generation of a subset of antigen-experienced $CD8^+$ T cells that reside in the barrier tissues or circulate between tissue, blood, and the secondary lymphoid organs to build a first line of defense against the reoccurrence of syngeneic tumors. Of note, is still disputed whether those memory $CD8^+$ T cells are differentiated from a subset of effector cells or from a distinct type of precursors (Enamorado et al., 2018).

DCs also activate additional types of lymphoid cells such as nature killer (NK) and B cells, which play critical roles in antitumor immunity. Mature DCs can recruit NK cells into tumor draining lymph nodes where they interact and form a positive feedback regulation loop. Upon engagement with NK cells, DCs produce IL-12, IL-15, and type I IFNs that stimulate the proliferation and activation of NK cells (Harizi, 2013; Munz et al., 2005), which in turn can further infiltrate the tumor bed and kill cancer cells through direct IFNγ-mediated cytolytic activity (Mittal et al., 2017; Yu et al., 2001). Moreover, interaction with NK cells leads to an increase in tumor microenvironment (TME)-infiltrating DCs and further stimulates their maturation (Bottcher et al., 2018). Activation of naïve B cells and the generation of memory B cells can also be ignited by DCs, resulting in an increase in antibody production and secretion (Dubois et al., 1997), as well as the maintenance of B cell memory (Jego et al., 2005).

3.3 DC as tumoricidal effector cells

Beyond their role in eliciting cytotoxic T cell and NK cell immunity, DCs can act as effector cells directly delivering death signals to cancer cells or slowing down their proliferation. Observations that support the direct tumoricidal effects of DCs are largely obtained from studies on myeloid DCs, which are endowed with the ability to mediate TNF-α-dependent tumor cell apoptosis (Fanger et al., 1999; Lu et al., 2002; Schmitz et al., 2005), antibody-mediated cellular cytotoxicity (Schmitz et al., 2002), as well as the generation of nitric oxide (Huang et al., 2005; Srivastava et al., 2007). Of note, the precise quantification of direct cancer cell killing by DCs is relatively complex. Potent cell-to-cell toxicity and soluble mediator-derived cytotoxicity are uniquely observed using ex vivo differentiated DCs, which are cocultured with cancer cell lines or freshly isolated primary tumor cells (Chapoval et al., 2000; Janjic et al., 2002). Additional studies employing freshly isolated primary DCs or even *in situ* monitoring of DC-cancer cell interactions should be performed to confirm direct DC-mediated anticancer effects.

The aforementioned mechanisms for DC-mediated antitumor immunity, though not covering all aspects of cellular interaction, have provided a solid rationale for the development of DC-based immunotherapies. As natural DCs constitute a relatively rare immune cell population, especially within tumors and lymphoid organs, autologous DC transfer offers the sole practical solution to harness the potential of DCs for cancer immunotherapies.

4. Concept of DC-based immunotherapies

At difference with adaptive T cell transfer therapy that aims at amplifying and accelerating direct cancer cell killing by selected or engineered effector T cells, DC-based immunotherapy intends to use appropriately activated DCs to induce and train the endogenous antitumor immune response, thus augmenting both effector and memory T cell responses for complete eradication of cancer cells and avoiding cancer recurrence for a prolonged period. The principle of DC-based immunotherapy is based on the extraction and ex vivo differentiation of autologous DCs from patients, their activation by loading with appropriate TAAs or exposure to whole tumor lysate, and finally their reinfusion to generate adaptive anticancer immunity. Considering the fact that adaptive DC transfer leads to a sort of immunization effect, it is generally referred to as therapeutic DC vaccination. For more than two decades, this approach has been evaluated in preclinical animal models as well as in clinical trials, emphasizing its potential to be developed into a safe and efficient therapeutic approach.

4.1 Generation of autologous DCs for adaptive transfer

Theoretically, there are two ways to load DCs with TAAs for efficient and stable antigen presentation to specific cells. The first is based on the direct in vivo delivery of antigens to naïve DCs in the host. By coupling TAAs with antibodies targeting specific DC surface markers, precise in vivo antigen delivery and enhanced cross presentation has been realized in animal models (Bonifaz et al., 2004; Hawiger et al., 2001), and safety and efficacy were further verified in clinical trials (Dhodapkar et al., 2014; Gargett et al., 2018). This approach avoids the complex ex vivo generation of DC-based vaccines, but the number of DCs that can be primed in vivo is very low, and systemic administration of antibodies or antigen complex introduce the risk of immune tolerance (Bonifaz et al., 2002; Hawiger et al., 2001). Actually, the majority of studies on DC vaccination, either preclinical or clinical, are based on ex vivo differentiated DCs or freshly isolated natural DCs that are pulsed with tumor associated antigens. The concept of loading tumor-associated antigens onto ex vivo differentiated DCs and reinfusing them back to the host to induce an antigen-specific immune response in vivo has been established about 30 years ago (Dhodapkar et al., 1999). Since then, continuous efforts have been devoted to generate DCs ex vivo in large numbers for vaccination studies.

In most clinical DC vaccination trials conducted so far, sufficient numbers of DCs were extracted from peripheral blood or ex vivo differentiated from hematopoietic precursors in the presence of specific cytokines. Typical examples using blood-extracted autologous DCs for adaptive transfer include Sipuleucel-T (commercial name Provenge), the first FDA-approved cell-based therapy (Kantoff et al., 2010), based on a series of clinical trials (Higano et al., 2009; Small et al., 2000, 2006). Of note, Sipuleucel-T is still the sole clinically-approved DC-based vaccine therapy. It consists of a proportion of $CD54^+$ DCs and other APCs that are extracted from peripheral blood mononuclear cells (PBMCs) and are pulsed with recombinant prostatic acid phosphatase (PAP) and GM-CSF (Kantoff et al., 2010). The most commonly used approach to prepare large numbers of autologous monocyte-derived DCs (moDC) consists in the differentiation of monocytes obtained from whole blood or leukapheresis. Administration of autologous MoDCs derived from $CD14^+$ peripheral monocytes or differentiated from $CD34^+$ progenitors is well tolerated and occasionally effective against cancer (Mastelic-Gavillet et al., 2019). Other routes to obtain DCs such as purification of naturally circulating blood DCs, allogenic DCs from healthy people or cancer patients, and DC-derived exosomes has also been tested as promising alternatives. A summary of such DC-centric trials is provided in Table 1.

Table 1 DC preparation for adaptive transfer in clinical trials.

Method	Source	Cytokines/PRR ligand	Antigen/stimulation	Reference/trials	Notes
Ex vivo culture	Peripheral blood	GM-CSF + IL-4	Tumor lysates	NCT01006044 (Geskin et al., 2018; Inoges et al., 2017; Redman et al., 2008)	CD14$^+$ monocytes selected from PBMCs as DC progenitors
		GM-CSF + IFN-γ	Tumor lysates	Baek et al. (2011)	CD34$^+$ HSCs from patient PBMCs as DC progenitors
		GM-CSF + FLT-3 + TNF-α + TGFβ1 or IL-4	Synthetic peptide	NCT00700167 (Romano et al., 2011; Ratzinger et al., 2004)	Langerhans cells or moDCs differentiated from CD34$^+$ HSCs
		GM-CSF + IL-4	Tumor mRNA + cytokines	NCT01278940 Kyte et al. (2006) NCT01983748	MoDCs differentiated from T and B cell-depleted PBMCs
		IFN-β + IL-3 + Poly(I:C)	Synthetic peptide	Buelens et al. (2002); Trakatelli et al. (2006)	DCs derived from PBMCs
	Umbilical cord blood	GM-CSF + IL-4	Cytokine cocktail, LPS, Poly(I:C)	Bedke et al. (2020); Plantinga et al. (2019)	Expansion from CD34$^+$ or CD14$^+$ progenitor cells from cord blood then differentiation to DCs
			None	NCT00731744	

	Source	Cytokines	Antigen	Reference	Description
Direct isolation	Peripheral blood	None	Recombinant TAAs/synthetic peptide	NCT02993315	Natural circulating CD1c$^+$ myeloid DCs and pDCs
		None	None	NCT03707808	Natural circulating CD1c$^+$ myeloid DCs
		None	Recombinant TAAs	NCT01133704	CD54$^+$ DCs and other APCs
Allogenic DCs	Peripheral blood	GM-CSF + IL-4	TNF-α	Florcken et al. (2013)	MoDCs from HLA-matched healthy PBMCs
	Sustainable DC progenitor	GM-CSF + TNF-α, + IL-4	Prostaglandin-E2, TNF-α, and IL-1β	NCT01373515 NCT03697707 van de Loosdrecht et al. (2018)	DCs generated from differentiated AML cell line DCOne with AML-associated antigens
DC–derived exosomes	Peripheral blood	GM-CSF + IL-4	Synthetic peptides and IFN-γ	Besse et al. (2016)	Exosomes extracted from antigen-loaded mature moDCs
			Synthetic peptides	Morse et al. (2005)	Antigen loaded onto exosomes extracted from immature moDCs

The optimal delivery of TAAs, as well as the stimulation of DC maturation, constitute further challenges. Due to the high abundance of TAAs and the reproducibility of the procedure, autologous or allogeneic tumor lysate and inactivated malignant cells are the most commonly used sources of antigens for DC priming. However, the preparation of tumor lysate or inactivated cancer cells requires the availability of a considerable quantity of fresh tumor specimen, which limits its employment to a few tumor types. The abundance of nonspecific antigens in tumor cell lysate may also saturate the antigen capture and presentation capability of DCs thus compromising the initiation of tumor-specific T cell responses. To overcome these hurdles, the use of recombinant or synthetic TAAs has become an attractive strategy that is tested in many clinical trials (Table 1). Synthetic TAAs can be designed to resemble certain cancer types, such as melanoma (Schreibelt et al., 2016; Tel et al., 2013) and prostate cancer (Kantoff et al., 2010), to ensure and improve tumor-specific T cell responses. Nevertheless, in some cases, synthetic TAAs may not exactly mimic the naturally processed epitopes because of the absence of posttranslational modifications (Valmori and Ayyoub, 2004). Transfecting or transducing autologous DCs with TAA-coding mRNA, cDNA or viral vectors, has been shown to induce a broader immune response than pulsing them with synthetic peptides or proteins because this strategy allows for optimal posttranslational processing of TAAs that are presented as natural epitopes (Breckpot et al., 2004; Butterfield et al., 2008; Kyte et al., 2006; Vik-Mo et al., 2013). Of note, numerous clinical trials use ex vivo differentiated DCs or purified natural DCs that are not exposed to antigen but rather are activated and matured in variable cocktails of cytokines and PRR agonists before they are reintroduced into patients (Bedke et al., 2020; Florcken et al., 2013; Plantinga et al., 2019; van de Loosdrecht et al., 2018). Obviously, the maturation stimuli and TAAs need to be carefully selected for targeting the appropriate DC subset and cancer type, respectively.

4.2 Preclinical model cell lines for cancer-relevant DC studies

DCs generated from human and murine bone marrow hematopoietic progenitors share a series of characteristics. Indeed, in preclinical experimentation, murine bone marrow derived DCs (BMDC) are the most commonly used DCs. Large quantities of immature or mature DCs can be derived from murine bone marrow derived hematopoietic precursors cultured with recombinant granulocyte-macrophage colony-stimulating factor (rGM-CSF). Of note, whether these BMDCs should be regarded as conventional DCs or moDCs, is still controversial. (Helft et al., 2015; Lutz et al., 1999, 2017) Recombinant FMS-like tyrosine kinase 3 ligand (FLT3L) is widely used to differentiate murine BMDCs that resemble pDCs, and a combination of FLT3L plus rGM-CSF can be used to generate BMDCs that resemble cDCs (Bjorck, 2001; van de Laar and Lambrecht, 2014). The number of cells that can be obtained via ex vivo differentiation is still relatively low, and it remains difficult to manipulate these terminally differentiated (and hence post-mitotic) primary cells, especially in terms of genetic modification. These disadvantages limit the utility of BMDCs for in vivo studies. As most laboratory mice are inbred strains, individuals are nearly identical in genotype thus becoming isogenic with respect to their MHC, allowing for the transfer of immune cells between mice from the same strain. This allows for the establishment of DC cell lines for long-term culture, their expansion and their use for genetic and functional studies. Theoretically, DC precursor cell lines can be infinitely expanded and injected into mice bearing the same MHC haplotype, for instance for the induction of anticancer immune responses. However, several widely used dendritic cell lines are not ideal for such in vivo applications. The DC2.4 cell line was established by transducing bone marrow cells with retroviral vectors expressing murine GM-CSF and super transfection with the *Myc* and *Raf* oncogenes (Shen et al., 1997). DC2.4 are homogeneous immature DC-like cells that can be easily cultured without the requirement of additional growth factors. This cell line has been widely used for functional and biochemical DC studies, especially in phagocytosis and antigen cross-presentation assays (Lee et al., 2004; Okada et al., 2001; Wang et al., 2020). However, DC2.4 is an established immortal cell line and has a strong potential to form tumors or develop into DC leukemia upon its injection into mice. MutuDCs originate from CD8α^+ cDC contained in splenic tumors from transgenic C57BL/6 mice expressing the SV40LgT oncogene under the CD11c promoter (Fuertes Marraco et al., 2012). This cell line has characteristics close to normal splenic cDC1 and can be used in cross-presentation assays *in vitro*. Protocols for establishing such mutuDCs from different transgenic mice are well established, and these DCs are also applicable for lentiviral transduction, making it a potentially useful tool for genotype-phenotype studies, yet again with the caveat of their malignant origin and their capacity to form tumors when administered to histocompatible mice (Fuertes Marraco et al., 2012). In an analogous fashion, human DC cell lines can be generated from the blood of patients with acute myeloid leukemia

(Kremser et al., 2010; Mohty et al., 2003). Although such leukemic DCs are valuable tools for studying DC biology in vitro, they should not be injected into patients.

Inducible immortalized immature DCs (iniDCs) may constitute a valid strategy for in vivo applications. IniDCs are differentiated from the bone marrow of an immorto-mouse expressing a tetracycline inducible simian virus (SV40) large T antigen (SV40LgT) (Richter et al., 2013). In the presence of doxycycline (DOX), the SV40LgT is activated to block retinoblastoma (RB) and tumor protein 53 (TP53), thus rendering the cells immortal (Anastassiadis et al., 2010; Richter et al., 2013). Moreover, the synthetic glucocorticoid dexamethasone (DEX) is added to cell culture medium to suppress the maturation/activation of iniDCs. IniDCs proliferate rapidly yet can be de-immortalized by the simultaneous removal of both DEX and DOX (Richter et al., 2013). This de-immortalization protocol avoids potential tumorigenesis in vivo, rendering iniDCs useful for the preclinical optimization of DC transfer-based immunotherapies. More importantly, it is feasible to genetically modify these iniDCs as they are infinitely expandable in the presence of DEX/DOX, and then to transform them into de-immortalized iniDCs (de-iniDCs) by the removal of DEX/DOX for phenotypic and functional studies in vitro and in vivo (Zhao et al., 2021).

5. Experimental and clinical approaches to improve DC-based immunotherapies

Since the first clinical evaluation of DC vaccines in 1996 (Hsu et al., 1996), DC transfer-based immunotherapy has been continuously optimized. The safety of DC transfer has been evaluated in numerous phase I and II clinical trials, showing that side effects are generally mild and transient as compared to other cell-based immunotherapies. Moreover, in most of these clinical trials, the induction of tumor-specific T cell immune response was documented. For example in prostate cancer, DC vaccination-induced tumor-specific immune responses in 77% of vaccinated patients (Draube et al., 2011). However, the clinical efficacy of DC transfer-based immunotherapies has been deceptive. According to a comprehensive meta-analysis, the objective response rate of prostate cancer patients to DC-based vaccination is only 7.1% (Anguille et al., 2014), meaning that less than 1 out of 10 patients with confirmed tumor-specific immune responses truly manifested shrinkage of the prostate cancer (Draube et al., 2011).

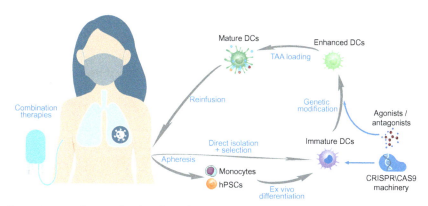

Fig. 1 A general principle of DC-based immunotherapy and possible routes to optimize the efficacy. Different subtypes of immature DCs can be obtained by either direct isolation based on bead selection, or ex vivo differentiation from monocytes or pluripotent stem cells (PSCs) through leukapheresis. Immature DCs can be subjected to CRISPR-mediated gene editing or pretreated by specific agonists or antagonists for generating phenotypes with increased potency or improving DC functions for initiating anticancer immune response, respectively. Such gene-engineered DCs can be loaded with appropriate tumor-associated antigens (TAA), and the "trained" mature DCs are then adaptively reinfused into patients. DC-based therapy can be further potentiated by combination with traditional chemotherapy, targeted therapies, as well as with additional immunotherapies.

Reasons for this discrepancy might lie in (i) the insufficient potency of the DC-mediated tumor-specific immune responses, (ii) an interference by immunosuppressive factors, and (iii) a shortage or exhaustion of downstream effector cells. As a result, most trials are now combining DC vaccines with other therapeutic approaches to overcome these shortcomings (Fig. 1).

5.1 Combination with adjuvants to potentiate DC-mediated immune responses

Immunostimulatory cytokines and TLR agonists have been applied to DC cultures or have been systemically administered to patients together with DC cell vaccines to boost immune priming. In this context, cytokines such as IL-2 has been proven to enhance vaccination efficacy in preclinical studies (Shimizu et al., 1999, 2000), but largely failed to boost antitumor effects in clinical trials (Baek et al., 2011; Miki et al., 2014). TLR agonists, such as poly(I:C), imiquimod or unmethylated CpG oligodeoxynucleotides, constitute another class of DC vaccine adjuvants that have achieved some success in boosting the efficacy of DC transfer-based therapies. Combining DC vaccination with an TLR3 agonist plus synthetic antigen and CD40 agonistic antibodies in preclinical models achieved successful remission of large subcutaneous murine MC38 tumors in immunocompetent animals and induced robust antigen-specific T-cell responses (Nimanong et al., 2017). In addition, TLR3 stimulation has been reported to improve anti-tumor immunity elicited by vaccination with DC-derived exosomes in a murine model of melanoma (Damo et al., 2015). Moreover, promising results without severe toxicity were obtained in clinical trials enrolling patients with recurrent malignant glioma (Okada et al., 2011), rhabdomyosarcoma (Salazar et al., 2014) and B-cell lymphoma (Brody et al., 2010).

The induction of ICD is emerging as an important strategy for the initiation of adaptive anticancer immune responses. ICD is characterized by the emission of adjuvant damage-associated molecular patterns (DAMPs) from dying cancer cells (Fucikova et al., 2020; Kroemer et al., 2013). DAMPs released in the course of ICD bind to specific PRRs expressed by DCs and initiate a cellular signal cascade that ultimately results in the activation of TAA-specific immune responses, supporting a potential synergistic effect between ICD inducing therapies and DC transfer-based immunotherapy, as well as the potential of ICD undergoing cancer cells as an adjuvant and a source of TAA to improve the efficacy of DC vaccination. Garg et al. combined DC immunotherapy with ICD induced by hypericin-based photodynamic therapy (PDT) in an orthotopic high-grade glioma (HGG) mouse model. This ICD-based DC vaccines provided strong anti-HGG effects and a significant survival benefit (Garg et al., 2016). A meta-analysis of the Cancer Genome Atlas glioblastoma cohort confirmed that this ICD-relevant immune signature is clinically associated with good patient prognosis, thus providing translation potential to ICD-based DC transfer therapies (Garg et al., 2016). Indeed, PDTs typically induce ICD in a variety of cancer types (Gomes-da-Silva et al., 2020), and PDT tumor lysate-pulsed DCs (PDT-DCs) inhibit the growth of mammary EMT6 tumors more potently than freeze/thaw tumor lysate-pulsed DC in mice, suggesting that PDT-DCs are particularly efficient (Jung et al., 2012). Similarly, splenic cDC1s pulsed with cancer cells undergoing irradiation-induced ICD were able to initiate therapeutic immune responses that synergized with checkpoint blockade (Wculek et al., 2019). This points to the possibility of advantageously combining DC-based therapies with immune checkpoint blockers targeting cytotoxic T-lymphocyte associated protein 4 (CTLA-4), programmed cell death 1 (PD-1) or its ligand PD-L1.

5.2 Exclusion of immunosuppressive factors to restore DC-mediated immune responses

A crucial factor limiting the success of DC transfer is the formation of an immunosuppressive TME, including the evasion of immune recognition by the loss of tumor antigen expression, reduced expression of MHC molecules and costimulatory markers, the production of immunosuppressive cytokines, as well as the recruitment of immunosuppressive regulatory cells (Hanahan and Weinberg, 2011). The depletion of regulatory T cells (Tregs) can break peripheral tolerance and may be particularly successful for improving DC transfer-based therapies. In murine breast cancer models, combining DC transfer with peptide P60, which targets the prototypic Treg transcription factor FOXP3, achieved superior antitumor efficacy compared to DC therapy alone (Moreno Ayala et al., 2017). In a murine glioma model, the elimination of Tregs using a CD25 neutralizing antibody was found essential for the therapeutic efficacy of tumor lysate-pulsed DCs (Grauer et al., 2008). Myeloid-derived suppressor cells (MDSCs) constitute a collection of immunosuppressive cells including macrophages and granulocytes that hamper the immune response and support tumor growth. Studies that focus on the specific depletion or inhibition of MDSCs are relatively rare, but a lot of immunomodulatory molecules have been reported to improve DC-based therapy by the reduction of MDSCs. For example, combination therapy with DCs and lenalidomide has been shown as an effective approach to enhance antitumor immunity in a mouse colon cancer model, correlating with the depletion of MDSCs (Vo et al., 2017). Inhibitors of the indoleamine 2,3-dioxygenase (IDO) pathway have been shown to inactivate MDSCs (as well as Tregs) and to improve an effective antitumor response (Prendergast et al., 2014). Clinical studies have evaluated the safety, immune and clinical effects of IDO-silenced DC vaccines, suggesting an enhanced immunogenic function of DCs in vitro and in vivo (Sioud et al., 2013). A phase II trial using sipuleucel-T followed by the IDO inhibitor indoximod also revealed a significant improvement of radiographic and clinical progression in patients with metastatic castration-resistant prostate cancer (Armstrong et al., 2019). Along with Tregs and MDCSs, tumor-associated macrophages (TAMs) may compromise anticancer immunity. Pharmacological inhibition of colony stimulating factor 1 receptor (CSF1R) with PLX3397 (or pexidartinib) depletes TAMs and synergizes with DC transfer therapy to eradicate mesothelioma in mouse models (Dammeijer et al., 2017).

5.3 In cooperation with other immunotherapies

The FDA approvals of ICBs targeting CTLA-4 or the PD-1/PD-L1 interaction have marked a major breakthrough in cancer immunotherapy. Theoretically, such ICBs may be combined with DC transfer to achieve unrestrained tumor-specific T-cell responses. Indeed, a number of studies combining DC transfer with ICBs have been performed or are currently ongoing. Enhanced DC transfer–induced T-cell responses were observed after CTLA-4 blockage, (Pierret et al., 2009; Ribas et al., 2005, 2009; Wilgenhof et al., 2016) and anti–CTLA-4 antibodies have even shown efficacy in metastatic melanoma patients who progressed after DC transfer (Boudewijns et al., 2016), further underscoring the potential of this combination strategy. DC vaccines combined with antibodies against PD-1 or PD-L1 showed improved therapeutic benefits as compared to DC transfer alone, leading to the eradication of murine lung cancer (Kadam and Sharma, 2020), as well as to enhanced survival of mice with metastatic HER2 breast cancer (Kodumudi et al., 2019). A pilot trial with 7 patients with stage IV pancreatic cancer that received systemic administration of anti-PD-1 together with DC transfer therapy achieved good responses, highlighting the potential of this combination strategy (Nesselhut et al., 2016). Of note, the same team has also demonstrated in an earlier trial that DC-induced T cell responses can be improved by direct blocking PD-L1 expressed on DCs (Nesselhut et al., 2015). More clinical trials employing this combination regime are now being conducted in different centers (NCT01420965; NCT01096602).

Chimeric antigen receptor T cells (CAR-T) have been extremely successful in the treatment of hematologic malignancies, yet showed limited efficacy against solid tumors. The classical "DC-cytotoxic T cell" antitumor immune response pathway has provided the rationale to design combinational therapies using DC vaccines and CAR-T cells. The first preclinical study based on this concept was published in 2018 (Akahori et al., 2018), showing that the efficacy of CAR-T cells expressing a single chain variable fragment (scFv) specific for Wilms tumor 1 (WT1) was further enhanced by vaccination with WT1-loaded DCs. This combination strategy is now under evaluation in a clinical trial aiming at assessing the safety and efficacy of CAR-T cells combined with peptide specific DCs for patients with relapsed/refractory leukemia (NCT03291444). In yet another preclinical setting, combination therapy with tumor-charged DCs and CAR-T cells exhibited cytotoxic effects against murine lymphoma and myeloma cells

(Capelletti et al., 2020). Moreover, several related clinical trials have shown that adoptive T cell therapy in combination with DC vaccination is safe, and that this combination regime can result in complete clinical response in patients with stage III/IV melanoma, who did not respond to ICBs (Lovgren et al., 2020; Poschke et al., 2014; Saberian et al., 2021; Wickström et al., 2018). Furthermore, the combination of TCR-engineered T cells together with DC transfer has proven to be safe and to cause at least transient antitumor activity (Chodon et al., 2014; Nowicki et al., 2019). Other more complex combination strategies that comprise more than 3 therapeutic approaches also achieved encouraging effects such as a randomized controlled phase III clinical trial combining chemotherapy, adoptive T cell transfer and DC transfer in patients with resected primary lung cancer. This trial showed extended overall survival of the group of patients receiving the triple therapy as compared to the groups receiving monotherapies (Kimura et al., 2015). In another phase II trial, metastatic melanoma patients were treated with autologous DCs as a vaccine, IL-2 and the cyclooxygenase-2 (Cox-2) inhibitor celecoxib as adjuvants, and cyclophosphamide as an immunogenic chemotherapy (Ellebaek et al., 2012). This triple combination resulted in twice as many patients obtaining stable disease with an average of a 6-month increase in survival as compared to a previous trial without cyclophosphamide and celecoxib (Trepiakas et al., 2010). Phase III trials are now needed to confirm these results.

6. Engineering dendritic cells to improve DC-based cancer immunotherapy

In vitro transcribed mRNA encoding tumor antigens has been extensively studied in preclinical models for the programming of DCs and showed promising results in clinical trials (Van Hoecke et al., 2021; Van Lint et al., 2016). Different from this procedure, a novel strategy for improving DC-based immunotherapy consists in the genetic engineering of DCs before their injection into oncological patients, thus creating DCs with improved anticancer activity. Such engineered DCs can carry knockouts of genes coding for checkpoint proteins or immunosuppressive cytokines such as TGF-β or, on the contrary overexpress transgenes coding for costimulatory molecules.

Current approaches for DC reprogramming include the RNAi-mediated silencing of immunosuppressive genes to trigger the onset of antigen-specific cytotoxic T cell immunity (Hobo et al., 2010; van der Waart et al., 2015), yet is hampered by technical hurdles that limit cellular viability. Furthermore, lentiviral and adenoviral transduction has been employed to overexpress key cytokines and chemokines in ex vivo differentiated moDCs and thus improve DC migration, lymph node homing (Okada et al., 2005) and infiltration of the tumor bed (Lee et al., 2017; Yang et al., 2004). However, viral transduction bears the risk of unwanted DC maturation thus rendering DCs incompetent of antigen uptake.

The optimization of the clustered regulatory interspaced short palindromic repeat (CRISPR)/CRISPR-associated protein 9 (Cas9) system offers a convenient technology for engineering DCs for adaptive transfer. Instead of using DCs from gene-edited transgenic mice (Han et al., 2019), primary DCs can be submitted to direct editing thanks to the novel CRISPR-CAS9 delivery technologies, such as the nucleofection of in vitro-assembled Cas9-sgRNA complexes (Freund et al., 2020; Jost et al., 2021), usually creating a pool composed by edited and non-edited DC, which then may be subjected to further characterization.

Obtaining large quantities of homologous gene-edited DCs for in vivo application is feasible by employing dendritic cell precursor cell lines such as iniDCs that are infinitely expandable and can be de-immortalized and differentiated into DC (Zhao et al., 2021). Genetic screens in iniDCs revealed loss- and gain of function phenotypes that might unveil actionable molecular targets for optimizing the immunotherapeutic performance of DCs. Gain-of-function in chemotaxis, synapse formation with dying cancer cells, phagocytosis, maturation and cross-presentation manifest provide the chance to engineer consummate antigen cross-presenting DCs ("CAR-DCs") with enhanced function. Furthermore, loss-of-function phenotypes may unveil immunosuppressive circuitries and yield heuristic insights into DC function (Fig. 1).

7. Concluding remarks

The concept of DC transfer-based immunotherapy has been established and developed based on the understanding of DC-induced antitumor immune responses. As the most potent antigen-presenting cell type, DCs are endowed with the ability to ignite and maintain tumor antigen-specific immune responses, leading to the priming of effector T-cells that recognize and eradicate cancer cells, while generating immunological memory which protects against tumor recurrence. In line with this hypothesis, cancer-specific T cell responses have been confirmed in almost all clinical trials evaluating DC transfer-based approaches as mono- or combination therapy. Compared with the more aggressive adaptive T cell transfer, DC therapy has been shown to be safe, while causing reversible mild-to-moderate side effects in a small proportion of patients. However, the anticancer efficacy achieved with DC transfer is still limited. Since the first clinical attempt to use DC transfer as a cancer therapy (Hsu et al., 1996), hundreds of trials have been reported. Nonetheless, Sipuleucel-T remains the only DC-based therapy that is FDA-approved for the treatment of metastatic castration-resistant prostate cancer (Kantoff et al., 2010). Of note, Sipuleucel-T is manufactured from autologous APC-containing PBMCs which also contain antigen-presenting immune cells other than DCs. Moreover, the commercial failure of Sipuleucel-T has had a negative impact on the further clinical development of DC transfer-based therapies.

Significant advances in optimizing DC-based immunotherapy over the past two decades have been achieved based on the detailed understanding of DC-mediated antitumor immune response. Earlier efforts focused on identifying the appropriate DC subtype and the strategies for isolating or differentiating these specific subsets that were believed to be most potent in antigen-cross presentation (Huber et al., 2018). The use of naturally circulating DCs entirely avoids the debate about the conditions of the ex vivo differentiation of DCs (Bol et al., 2019), but the quantity of DCs that can be obtained through this approach remains limited. More effort has been devoted to improving the activation and maturation of DCs, including the selection of tumor-associated antigens, the combination with adjuvants, and the systemic delivery of antigens (Sabado et al., 2017). These approaches mainly aim at optimizing DC transfer as a monotherapy. Recent clinical trials tend towards the combination of DC-transfer with the enforced (re-)activation of downstream effector T cells by ICBs or with CAR-T cells (Sprooten et al., 2019). CAR-T cells illustrate that genetically engineered cells can be remarkably successful in the clinics, especially in the treatment of hematological malignancies (Xin et al., 2019). This has inspired us to develop engineering strategies for the genetic exploration of anticancer-relevant DC functions, aiming at revealing more insights into our understanding of DC-mediated anticancer immune response. The perspective of exploiting systematic genetic screens with the aim of enhancing DC function for improved anti-tumor immunotherapy is encouraging and the validation of these findings in well-designed clinical trials will hopefully build the foundation for future DC-based therapies.

Acknowledgments

OK receives funding by the DIM ELICIT initiative of the Ile de France and Institut National du Cancer (INCa); GK are supported by the Ligue contre le Cancer (équipes labellisées, Program "Equipe labelisée LIGUE"; no. EL2016.LNCC (VT/PLP)); Agence National de la Recherche (ANR) – Projets blancs; ANR under the frame of E-Rare-2, the ERA-Net for Research on Rare Diseases; AMMICa US23/CNRS UMS3655; Association pour la recherche sur le cancer (ARC); Association "Le Cancer du Sein, Parlons-en!"; Cancéropôle Ile-de-France; Chancelerie des universités de Paris (Legs Poix), Fondation pour la Recherche Médicale (FRM); a donation by Elior; European Research Area Network on Cardiovascular Diseases (ERA-CVD, MINOTAUR); Gustave Roussy Odyssea, the European Union Horizon 2020 Project Oncobiome; Fondation Carrefour; INCa; Inserm (HTE); Institut Universitaire de France; LeDucq Foundation; the LabEx Immuno-Oncology (ANR-18-IDEX-0001); the RHU Torino Lumière; the Seerave Foundation; the SIRIC Stratified Oncology Cell DNA Repair and Tumor Immune Elimination (SOCRATE); and the SIRIC Cancer Research and Personalized Medicine (CARPEM).

Conflict of interest

GK and OK are cofounders of Samsara Therapeutics. GK is a cofounder of everImmune and Therafast Bio.

References

Aarntzen, E.H., et al., 2012. Skin-test infiltrating lymphocytes early predict clinical outcome of dendritic cell-based vaccination in metastatic melanoma. Cancer Res. 72, 6102–6110. https://doi.org/10.1158/0008-5472.CAN-12-2479.

Akahori, Y., et al., 2018. Antitumor activity of CAR-T cells targeting the intracellular oncoprotein WT1 can be enhanced by vaccination. Blood 132, 1134–1145. https://doi.org/10.1182/blood-2017-08-802926.

Albert, M.L., Sauter, B., Bhardwaj, N., 1998. Dendritic cells acquire antigen from apoptotic cells and induce class I-restricted CTLs. Nature 392, 86–89. https://doi.org/10.1038/32183.

Anastassiadis, K., Rostovskaya, M., Lubitz, S., Weidlich, S., Stewart, A.F., 2010. Precise conditional immortalization of mouse cells using tetracycline-regulated SV40 large T-antigen. Genesis 48, 220–232. https://doi.org/10.1002/dvg.20605.

Anguille, S., Smits, E.L., Lion, E., van Tendeloo, V.F., Berneman, Z.N., 2014. Clinical use of dendritic cells for cancer therapy. Lancet Oncol. 15, e257–e267. https://doi.org/10.1016/S1470-2045(13)70585-0.

Apetoh, L., et al., 2007. Toll-like receptor 4-dependent contribution of the immune system to anticancer chemotherapy and radiotherapy. Nat. Med. 13, 1050–1059. https://doi.org/10.1038/nm1622.

Armstrong, A.J., et al., 2019. ARCHES: a randomized, phase III study of androgen deprivation therapy with enzalutamide or placebo in men with metastatic hormone-sensitive prostate cancer. J. Clin. Oncol. 37, 2974–2986. https://doi.org/10.1200/JCO.19.00799.

Baek, S., et al., 2011. Combination therapy of renal cell carcinoma or breast cancer patients with dendritic cell vaccine and IL-2: results from a phase I/II trial. J. Transl. Med. 9, 178. https://doi.org/10.1186/1479-5876-9-178.

Bai, X.F., et al., 2001. On the site and mode of antigen presentation for the initiation of clonal expansion of CD8 T cells specific for a natural tumor antigen. Cancer Res. 61, 6860–6867.

Banchereau, J., et al., 2000. Immunobiology of dendritic cells. Annu. Rev. Immunol. 18, 767–811. https://doi.org/10.1146/annurev.immunol.18.1.767.

Basu, S., Binder, R.J., Ramalingam, T., Srivastava, P.K., 2001. CD91 is a common receptor for heat shock proteins gp96, hsp90, hsp70, and calreticulin. Immunity 14, 303–313. https://doi.org/10.1016/s1074-7613(01)00111-x.

Bedke, N., et al., 2020. A method for the generation of large numbers of dendritic cells from CD34+ hematopoietic stem cells from cord blood. J. Immunol. Methods 477, 112703. https://doi.org/10.1016/j.jim.2019.112703.

Besse, B., et al., 2016. Dendritic cell-derived exosomes as maintenance immunotherapy after first line chemotherapy in NSCLC. Oncoimmunology 5, e1071008. https://doi.org/10.1080/2162402X.2015.1071008.

Bjorck, P., 2001. Isolation and characterization of plasmacytoid dendritic cells from Flt3 ligand and granulocyte-macrophage colony-stimulating factor-treated mice. Blood 98, 3520–3526. https://doi.org/10.1182/blood.v98.13.3520.

Bol, K.F., et al., 2019. The clinical application of cancer immunotherapy based on naturally circulating dendritic cells. J. Immunother. Cancer 7, 109. https://doi.org/10.1186/s40425-019-0580-6.

Bonifaz, L., et al., 2002. Efficient targeting of protein antigen to the dendritic cell receptor DEC-205 in the steady state leads to antigen presentation on major histocompatibility complex class I products and peripheral CD8+ T cell tolerance. J. Exp. Med. 196, 1627–1638. https://doi.org/10.1084/jem.20021598.

Bonifaz, L.C., et al., 2004. In vivo targeting of antigens to maturing dendritic cells via the DEC-205 receptor improves T cell vaccination. J. Exp. Med. 199, 815–824. https://doi.org/10.1084/jem.20032220.

Bottcher, J.P., et al., 2018. NK cells stimulate recruitment of cDC1 into the tumor microenvironment promoting cancer immune control. Cell 172, 1022–1037. e1014. https://doi.org/10.1016/j.cell.2018.01.004.

Boudewijns, S., et al., 2016. Ipilimumab administered to metastatic melanoma patients who progressed after dendritic cell vaccination. Oncoimmunology 5, e1201625. https://doi.org/10.1080/2162402X.2016.1201625.

Breckpot, K., Heirman, C., De Greef, C., van der Bruggen, P., Thielemans, K., 2004. Identification of new antigenic peptide presented by HLA-Cw7 and encoded by several MAGE genes using dendritic cells transduced with lentiviruses. J. Immunol. 172, 2232–2237. https://doi.org/10.4049/jimmunol.172.4.2232.

Brody, J.D., et al., 2010. In situ vaccination with a TLR9 agonist induces systemic lymphoma regression: a phase I/II study. J. Clin. Oncol. 28, 4324–4332. https://doi.org/10.1200/JCO.2010.28.9793.

Buelens, C., et al., 2002. Interleukin-3 and interferon beta cooperate to induce differentiation of monocytes into dendritic cells with potent helper T-cell stimulatory properties. Blood 99, 993–998. https://doi.org/10.1182/blood.v99.3.993.

Butterfield, L.H., et al., 2008. Adenovirus MART-1-engineered autologous dendritic cell vaccine for metastatic melanoma. J. Immunother. 31, 294–309. https://doi.org/10.1097/CJI.0b013e31816a8910.

Capelletti, M., et al., 2020. Potent synergy between combination of chimeric antigen receptor (CAR) therapy targeting CD19 in conjunction with dendritic cell (DC)/tumor fusion vaccine in hematological malignancies. Biol. Blood Marrow Transplant. 26, S42–S43. https://doi.org/10.1016/j.bbmt.2019.12.110.

Chan, C.W., Housseau, F., 2008. The 'kiss of death' by dendritic cells to cancer cells. Cell Death Differ. 15, 58–69. https://doi.org/10.1038/sj.cdd.4402235.

Chan, T., et al., 2007. Enhanced T-cell immunity induced by dendritic cells with phagocytosis of heat shock protein 70 gene-transfected tumor cells in early phase of apoptosis. Cancer Gene Ther. 14, 409–420. https://doi.org/10.1038/sj.cgt.7701025.

Chapoval, A.I., Tamada, K., Chen, L., 2000. In vitro growth inhibition of a broad spectrum of tumor cell lines by activated human dendritic cells. Blood 95, 2346–2351.

Chodon, T., et al., 2014. Adoptive transfer of MART-1 T-cell receptor transgenic lymphocytes and dendritic cell vaccination in patients with metastatic melanoma. Clin. Cancer Res. 20, 2457–2465. https://doi.org/10.1158/1078-0432.CCR-13-3017.

Collin, M., Bigley, V., 2018. Human dendritic cell subsets: an update. Immunology 154, 3–20. https://doi.org/10.1111/imm.12888.

Dammeijer, F., et al., 2017. Depletion of tumor-associated macrophages with a CSF-1R kinase inhibitor enhances antitumor immunity and survival induced by DC immunotherapy. Cancer Immunol. Res. 5, 535–546. https://doi.org/10.1158/2326-6066.CIR-16-0309.

Damo, M., Wilson, D.S., Simeoni, E., Hubbell, J.A., 2015. TLR-3 stimulation improves anti-tumor immunity elicited by dendritic cell exosome-based vaccines in a murine model of melanoma. Sci. Rep. 5, 17622. https://doi.org/10.1038/srep17622.

de Jong, J.M., et al., 2006. Dendritic cells, but not macrophages or B cells, activate major histocompatibility complex class II-restricted CD4+ T cells upon immune-complex uptake in vivo. Immunology 119, 499–506. https://doi.org/10.1111/j.1365-2567.2006.02464.x.

Dhodapkar, M.V., et al., 1999. Rapid generation of broad T-cell immunity in humans after a single injection of mature dendritic cells. J. Clin. Invest. 104, 173–180. https://doi.org/10.1172/JCI6909.

Dhodapkar, M.V., et al., 2014. Induction of antigen-specific immunity with a vaccine targeting NY-ESO-1 to the dendritic cell receptor DEC-205. Sci. Transl. Med. 6, 232ra251. https://doi.org/10.1126/scitranslmed.3008068.

Diao, J., et al., 2006. In situ replication of immediate dendritic cell (DC) precursors contributes to conventional DC homeostasis in lymphoid tissue. J. Immunol. 176, 7196–7206. https://doi.org/10.4049/jimmunol.176.12.7196.

Draube, A., et al., 2011. Dendritic cell based tumor vaccination in prostate and renal cell cancer: a systematic review and meta-analysis. PLoS One 6, e18801. https://doi.org/10.1371/journal.pone.0018801.

Dubois, B., et al., 1997. Dendritic cells enhance growth and differentiation of CD40-activated B lymphocytes. J. Exp. Med. 185, 941–951. https://doi.org/10.1084/jem.185.5.941.

Ellebaek, E., et al., 2012. Metastatic melanoma patients treated with dendritic cell vaccination, Interleukin-2 and metronomic cyclophosphamide: results from a phase II trial. Cancer Immunol. Immunother. 61, 1791–1804. https://doi.org/10.1007/s00262-012-1242-4.

Enamorado, M., Khouili, S.C., Iborra, S., Sancho, D., 2018. Genealogy, dendritic cell priming, and differentiation of tissue-resident memory CD8(+) T cells. Front. Immunol. 9, 1751. https://doi.org/10.3389/fimmu.2018.01751.

Fanger, N.A., Maliszewski, C.R., Schooley, K., Griffith, T.S., 1999. Human dendritic cells mediate cellular apoptosis via tumor necrosis factor-related apoptosis-inducing ligand (TRAIL). J. Exp. Med. 190, 1155–1164. https://doi.org/10.1084/jem.190.8.1155.

Farhood, B., Najafi, M., Mortezaee, K., 2019. CD8(+) cytotoxic T lymphocytes in cancer immunotherapy: a review. J. Cell. Physiol. 234, 8509–8521. https://doi.org/10.1002/jcp.27782.

Florcken, A., et al., 2013. Allogeneic partially HLA-matched dendritic cells pulsed with autologous tumor cell lysate as a vaccine in metastatic renal cell cancer: a clinical phase I/II study. Hum. Vaccin. Immunother. 9, 1217–1227. https://doi.org/10.4161/hv.24149.

Freund, E.C., et al., 2020. Efficient gene knockout in primary human and murine myeloid cells by non-viral delivery of CRISPR-Cas9. J. Exp. Med. 217. https://doi.org/10.1084/jem.20191692.

Fucikova, J., et al., 2020. Detection of immunogenic cell death and its relevance for cancer therapy. Cell Death Dis. 11, 1013. https://doi.org/10.1038/s41419-020-03221-2.

Fuertes Marraco, S.A., et al., 2012. Novel murine dendritic cell lines: a powerful auxiliary tool for dendritic cell research. Front. Immunol. 3, 331. https://doi.org/10.3389/fimmu.2012.00331.

Fujii, S., Shimizu, K., Kronenberg, M., Steinman, R.M., 2002. Prolonged IFN-gamma-producing NKT response induced with alpha-galactosylceramide-loaded DCs. Nat. Immunol. 3, 867–874. https://doi.org/10.1038/ni827.

Garg, A.D., et al., 2016. Dendritic cell vaccines based on immunogenic cell death elicit danger signals and T cell-driven rejection of high-grade glioma. Sci. Transl. Med. 8, 328ra327. https://doi.org/10.1126/scitranslmed.aae0105.

Gargett, T., et al., 2018. Phase I trial of Lipovaxin-MM, a novel dendritic cell-targeted liposomal vaccine for malignant melanoma. Cancer Immunol. Immunother. 67, 1461–1472. https://doi.org/10.1007/s00262-018-2207-z.

Geijtenbeek, T.B., van Vliet, S.J., Engering, A., 't Hart, B.A., van Kooyk, Y., 2004. Self- and nonself-recognition by C-type lectins on dendritic cells. Annu. Rev. Immunol. 22, 33–54. https://doi.org/10.1146/annurev.immunol.22.012703.104558.

Geskin, L.J., et al., 2018. Three antigen-loading methods in dendritic cell vaccines for metastatic melanoma. Melanoma Res. 28, 211–221. https://doi.org/10.1097/CMR.0000000000000441.

Ghiringhelli, F., et al., 2005. Tumor cells convert immature myeloid dendritic cells into TGF-beta-secreting cells inducing CD4+CD25+ regulatory T cell proliferation. J. Exp. Med. 202, 919–929. https://doi.org/10.1084/jem.20050463.

Ghiringhelli, F., et al., 2009. Activation of the NLRP3 inflammasome in dendritic cells induces IL-1beta-dependent adaptive immunity against tumors. Nat. Med. 15, 1170–1178. https://doi.org/10.1038/nm.2028.

Gomes-da-Silva, L.C., Kepp, O., Kroemer, G., 2020. Regulatory approval of photo-immunotherapy: photodynamic therapy that induces immunogenic cell death. Oncoimmunology 9, 1841393. https://doi.org/10.1080/2162402X.2020.1841393.

Grauer, O.M., et al., 2008. Elimination of regulatory T cells is essential for an effective vaccination with tumor lysate-pulsed dendritic cells in a murine glioma model. Int. J. Cancer 122, 1794–1802. https://doi.org/10.1002/ijc.23284.

Grouard, G., et al., 1997. The enigmatic plasmacytoid T cells develop into dendritic cells with interleukin (IL)-3 and CD40-ligand. J. Exp. Med. 185, 1101–1111. https://doi.org/10.1084/jem.185.6.1101.

Guermonprez, P., Valladeau, J., Zitvogel, L., Thery, C., Amigorena, S., 2002. Antigen presentation and T cell stimulation by dendritic cells. Annu. Rev. Immunol. 20, 621–667. https://doi.org/10.1146/annurev.immunol.20.100301.064828.

Han, D., et al., 2019. Anti-tumour immunity controlled through mRNA m(6)A methylation and YTHDF1 in dendritic cells. Nature 566, 270–274. https://doi.org/10.1038/s41586-019-0916-x.

Hanahan, D., Weinberg, R.A., 2011. Hallmarks of cancer: the next generation. Cell 144, 646–674. https://doi.org/10.1016/j.cell.2011.02.013.

Harizi, H., 2013. Reciprocal crosstalk between dendritic cells and natural killer cells under the effects of PGE2 in immunity and immunopathology. Cell. Mol. Immunol. 10, 213–221. https://doi.org/10.1038/cmi.2013.1.

Hawiger, D., et al., 2001. Dendritic cells induce peripheral T cell unresponsiveness under steady state conditions in vivo. J. Exp. Med. 194, 769–779. https://doi.org/10.1084/jem.194.6.769.

Helft, J., et al., 2015. GM-CSF mouse bone marrow cultures comprise a heterogeneous population of CD11c(+)MHCII(+) macrophages and dendritic cells. Immunity 42, 1197–1211. https://doi.org/10.1016/j.immuni.2015.05.018.

Heufler, C., et al., 1996. Interleukin-12 is produced by dendritic cells and mediates T helper 1 development as well as interferon-gamma production by T helper 1 cells. Eur. J. Immunol. 26, 659–668. https://doi.org/10.1002/eji.1830260323.

Higano, C.S., et al., 2009. Integrated data from 2 randomized, double-blind, placebo-controlled, phase 3 trials of active cellular immunotherapy with sipuleucel-T in advanced prostate cancer. Cancer 115, 3670–3679. https://doi.org/10.1002/cncr.24429.

Hobo, W., et al., 2010. siRNA silencing of PD-L1 and PD-L2 on dendritic cells augments expansion and function of minor histocompatibility antigen-specific CD8+ T cells. Blood 116, 4501–4511. https://doi.org/10.1182/blood-2010-04-278739.

Hsu, F.J., et al., 1996. Vaccination of patients with B-cell lymphoma using autologous antigen-pulsed dendritic cells. Nat. Med. 2, 52–58. https://doi.org/10.1038/nm0196-52.

Huang, J., Tatsumi, T., Pizzoferrato, E., Vujanovic, N., Storkus, W.J., 2005. Nitric oxide sensitizes tumor cells to dendritic cell-mediated apoptosis, uptake, and cross-presentation. Cancer Res. 65, 8461–8470. https://doi.org/10.1158/0008-5472.CAN-05-0654.

Huber, A., Dammeijer, F., Aerts, J., Vroman, H., 2018. Current state of dendritic cell-based immunotherapy: opportunities for in vitro antigen loading of different DC subsets? Front. Immunol. 9, 2804. https://doi.org/10.3389/fimmu.2018.02804.

Inoges, S., et al., 2017. A phase II trial of autologous dendritic cell vaccination and radio-chemotherapy following fluorescence-guided surgery in newly diagnosed glioblastoma patients. J. Transl. Med. 15, 104. https://doi.org/10.1186/s12967-017-1202-z.

Ishigami, S., et al., 2000. Clinical impact of intratumoral natural killer cell and dendritic cell infiltration in gastric cancer. Cancer Lett. 159, 103–108. https://doi.org/10.1016/s0304-3835(00)00542-5.

Iyoda, T., et al., 2002. The CD8+ dendritic cell subset selectively endocytoses dying cells in culture and in vivo. J. Exp. Med. 195, 1289–1302. https://doi.org/10.1084/jem.20020161.

Janjic, B.M., et al., 2002. Innate direct anticancer effector function of human immature dendritic cells. I. Involvement of an apoptosis-inducing pathway. J. Immunol. 168, 1823–1830. https://doi.org/10.4049/jimmunol.168.4.1823.

Jego, G., Pascual, V., Palucka, A.K., Banchereau, J., 2005. Dendritic cells control B cell growth and differentiation. Curr. Dir. Autoimmun. 8, 124–139. https://doi.org/10.1159/000082101.

Jost, M., et al., 2021. CRISPR-based functional genomics in human dendritic cells. Elife 10. https://doi.org/10.7554/eLife.65856.

Jung, N.C., et al., 2012. Photodynamic therapy-mediated DC immunotherapy is highly effective for the inhibition of established solid tumors. Cancer Lett. 324, 58–65. https://doi.org/10.1016/j.canlet.2012.04.024.

Kadam, P., Sharma, S., 2020. PD-1 immune checkpoint blockade promotes therapeutic cancer vaccine to eradicate lung cancer. Vaccines (Basel) 8. https://doi.org/10.3390/vaccines8020317.

Kambayashi, T., Laufer, T.M., 2014. Atypical MHC class II-expressing antigen-presenting cells: can anything replace a dendritic cell? Nat. Rev. Immunol. 14, 719–730. https://doi.org/10.1038/nri3754.

Kantoff, P.W., et al., 2010. Sipuleucel-T immunotherapy for castration-resistant prostate cancer. N. Engl. J. Med. 363, 411–422. https://doi.org/10.1056/NEJMoa1001294.

Kimura, H., et al., 2015. Randomized controlled phase III trial of adjuvant chemo-immunotherapy with activated killer T cells and dendritic cells in patients with resected primary lung cancer. Cancer Immunol. Immunother. 64, 51–59. https://doi.org/10.1007/s00262-014-1613-0.

Kodumudi, K.N., et al., 2019. Sequential anti-PD1 therapy following dendritic cell vaccination improves survival in a HER2 mammary carcinoma model and identifies a critical role for CD4 T cells in mediating the response. Front. Immunol. 10, 1939. https://doi.org/10.3389/fimmu.2019.01939.

Koucky, V., Boucek, J., Fialova, A., 2019. Immunology of plasmacytoid dendritic cells in solid tumors: a brief review. Cancers (Basel) 11. https://doi.org/10.3390/cancers11040470.

Kremser, A., et al., 2010. Dendritic cells (DCs) can be successfully generated from leukemic blasts in individual patients with AML or MDS: an evaluation of different methods. J. Immunother. 33, 185–199. https://doi.org/10.1097/CJI.0b013e3181b8f4ce.

Kroemer, G., Galluzzi, L., Kepp, O., Zitvogel, L., 2013. Immunogenic cell death in cancer therapy. Annu. Rev. Immunol. 31, 51–72. https://doi.org/10.1146/annurev-immunol-032712-100008.

Kyte, J.A., et al., 2006. Phase I/II trial of melanoma therapy with dendritic cells transfected with autologous tumor-mRNA. Cancer Gene Ther. 13, 905–918. https://doi.org/10.1038/sj.cgt.7700961.

Lee, H.Y., et al., 2004. Trp-Lys-Tyr-Met-Val-Met stimulates phagocytosis via phospholipase D-dependent signaling in mouse dendritic cells. Exp. Mol. Med. 36, 135–144. https://doi.org/10.1038/emm.2004.20.

Lee, J.M., et al., 2017. Phase I trial of intratumoral injection of CCL21 gene-modified dendritic cells in lung cancer elicits tumor-specific immune responses and CD8(+) T-cell infiltration. Clin. Cancer Res. 23, 4556–4568. https://doi.org/10.1158/1078-0432.CCR-16-2821.

Levin, D., Constant, S., Pasqualini, T., Flavell, R., Bottomly, K., 1993. Role of dendritic cells in the priming of CD4+ T lymphocytes to peptide antigen in vivo. J. Immunol. 151, 6742–6750.

Liu, J., Zhang, X., Cheng, Y., Cao, X., 2021. Dendritic cell migration in inflammation and immunity. Cell. Mol. Immunol. 18, 2461–2471. https://doi.org/10.1038/s41423-021-00726-4.

Lovgren, T., et al., 2020. Complete and long-lasting clinical responses in immune checkpoint inhibitor-resistant, metastasized melanoma treated with adoptive T cell transfer combined with DC vaccination. Oncoimmunology 9, 1792058. https://doi.org/10.1080/2162402X.2020.1792058.

Lu, G., et al., 2002. Innate direct anticancer effector function of human immature dendritic cells. II. Role of TNF, lymphotoxin-alpha(1)beta(2), Fas ligand, and TNF-related apoptosis-inducing ligand. J. Immunol. 168, 1831–1839. https://doi.org/10.4049/jimmunol.168.4.1831.

Lutz, M.B., et al., 1999. An advanced culture method for generating large quantities of highly pure dendritic cells from mouse bone marrow. J. Immunol. Methods 223, 77–92. https://doi.org/10.1016/s0022-1759(98)00204-x.

Lutz, M.B., Strobl, H., Schuler, G., Romani, N., 2017. GM-CSF monocyte-derived cells and langerhans cells as part of the dendritic cell family. Front. Immunol. 8, 1388. https://doi.org/10.3389/fimmu.2017.01388.

Maldonado-Lopez, R., Moser, M., 2001. Dendritic cell subsets and the regulation of Th1/Th2 responses. Semin. Immunol. 13, 275–282. https://doi.org/10.1006/smim.2001.0323.

Mastelic-Gavillet, B., Balint, K., Boudousquie, C., Gannon, P.O., Kandalaft, L.E., 2019. Personalized dendritic cell vaccines-recent breakthroughs and encouraging clinical results. Front. Immunol. 10, 766. https://doi.org/10.3389/fimmu.2019.00766.

Merad, M., Sathe, P., Helft, J., Miller, J., Mortha, A., 2013. The dendritic cell lineage: ontogeny and function of dendritic cells and their subsets in the steady state and the inflamed setting. Annu. Rev. Immunol. 31, 563–604. https://doi.org/10.1146/annurev-immunol-020711-074950.

Miki, K., et al., 2014. Combination therapy with dendritic cell vaccine and IL-2 encapsulating polymeric micelles enhances intra-tumoral accumulation of antigen-specific CTLs. Int. Immunopharmacol. 23, 499–504. https://doi.org/10.1016/j.intimp.2014.09.025.

Mittal, D., et al., 2017. Interleukin-12 from CD103(+) Batf3-dependent dendritic cells required for NK-cell suppression of metastasis. Cancer Immunol. Res. 5, 1098–1108. https://doi.org/10.1158/2326-6066.CIR-17-0341.

Mohty, M., Gaugler, B., Olive, D., 2003. Generation of leukemic dendritic cells from patients with acute myeloid leukemia. Methods Mol. Biol. 215, 463–471. https://doi.org/10.1385/1-59259-345-3:463.

Moreno Ayala, M.A., et al., 2017. Therapeutic blockade of Foxp3 in experimental breast cancer models. Breast Cancer Res. Treat 166, 393–405. https://doi.org/10.1007/s10549-017-4414-2.

Morse, M.A., et al., 2005. A phase I study of dexosome immunotherapy in patients with advanced non-small cell lung cancer. J. Transl. Med. 3, 9. https://doi.org/10.1186/1479-5876-3-9.

Munz, C., et al., 2005. Mature myeloid dendritic cell subsets have distinct roles for activation and viability of circulating human natural killer cells. Blood 105, 266–273. https://doi.org/10.1182/blood-2004-06-2492.

Naik, S.H., et al., 2006. Intrasplenic steady-state dendritic cell precursors that are distinct from monocytes. Nat. Immunol. 7, 663–671. https://doi.org/10.1038/ni1340.

Nesselhut, J., et al., 2015. Dendritic cells generated with PDL-1 checkpoint blockade for treatment of advanced pancreatic cancer. J. Clin. Oncol. 33, 4128. https://doi.org/10.1200/jco.2015.33.15_suppl.4128.

Nesselhut, J., et al., 2016. Systemic treatment with anti-PD-1 antibody nivolumab in combination with vaccine therapy in advanced pancreatic cancer. J. Clin. Oncol. 34, 3092. https://doi.org/10.1200/JCO.2016.34.15_suppl.3092.

Nimanong, S., et al., 2017. CD40 signaling drives potent cellular immune responses in heterologous cancer vaccinations. Cancer Res. 77, 1918–1926. https://doi.org/10.1158/0008-5472.CAN-16-2089.

Nowicki, T.S., et al., 2019. A pilot trial of the combination of transgenic NY-ESO-1-reactive adoptive cellular therapy with dendritic cell vaccination with or without ipilimumab. Clin. Cancer Res. 25, 2096–2108. https://doi.org/10.1158/1078-0432.CCR-18-3496.

Obeid, M., et al., 2007. Calreticulin exposure dictates the immunogenicity of cancer cell death. Nat. Med. 13, 54–61. https://doi.org/10.1038/nm1523.

Okada, N., et al., 2001. Administration route-dependent vaccine efficiency of murine dendritic cells pulsed with antigens. Br. J. Cancer 84, 1564–1570. https://doi.org/10.1054/bjoc.2001.1801.

Okada, N., et al., 2005. Augmentation of the migratory ability of DC-based vaccine into regional lymph nodes by efficient CCR7 gene transduction. Gene Ther. 12, 129–139. https://doi.org/10.1038/sj.gt.3302358.

Okada, H., et al., 2011. Induction of CD8 + T-cell responses against novel glioma-associated antigen peptides and clinical activity by vaccinations with {alpha}-type 1 polarized dendritic cells and polyinosinic-polycytidylic acid stabilized by lysine and carboxymethylcellulose in patients with recurrent malignant glioma. J. Clin. Oncol. 29, 330–336. https://doi.org/10.1200/JCO.2010.30.7744.

Paglia, P., Chiodoni, C., Rodolfo, M., Colombo, M.P., 1996. Murine dendritic cells loaded in vitro with soluble protein prime cytotoxic T lymphocytes against tumor antigen in vivo. J. Exp. Med. 183, 317–322. https://doi.org/10.1084/jem.183.1.317.

Park, S.Y., Kim, I.S., 2017. Engulfment signals and the phagocytic machinery for apoptotic cell clearance. Exp. Mol. Med. 49, e331. https://doi.org/10.1038/emm.2017.52.

Pham, T., et al., 2018. An update on immunotherapy for solid tumors: a review. Ann. Surg. Oncol. 25, 3404–3412. https://doi.org/10.1245/s10434-018-6658-4.

Pierret, L., et al., 2009. Correlation between prior therapeutic dendritic cell vaccination and the outcome of patients with metastatic melanoma treated with ipilimumab. J. Clin. Oncol. 27, e20006. https://doi.org/10.1200/jco.2009.27.15_suppl.e20006.

Plantinga, M., et al., 2019. Cord-blood-stem-cell-derived conventional dendritic cells specifically originate from CD115-expressing precursors. Cancers (Basel) 11. https://doi.org/10.3390/cancers11020181.

Platt, C.D., et al., 2010. Mature dendritic cells use endocytic receptors to capture and present antigens. Proc. Natl. Acad. Sci. U.S.A. 107, 4287–4292. https://doi.org/10.1073/pnas.0910609107.

Poschke, I., et al., 2014. A phase I clinical trial combining dendritic cell vaccination with adoptive T cell transfer in patients with stage IV melanoma. Cancer Immunol. Immunother. 63, 1061–1071. https://doi.org/10.1007/s00262-014-1575-2.

Prendergast, G.C., et al., 2014. Indoleamine 2,3-dioxygenase pathways of pathogenic inflammation and immune escape in cancer. Cancer Immunol. Immunother. 63, 721–735. https://doi.org/10.1007/s00262-014-1549-4.

Rabinovitch, M., 1995. Professional and non-professional phagocytes: an introduction. Trends Cell Biol. 5, 85–87. https://doi.org/10.1016/s0962-8924(00)88955-2.

Rao, Q., et al., 2016. Tumor-derived exosomes elicit tumor suppression in murine hepatocellular carcinoma models and humans in vitro. Hepatology 64, 456–472. https://doi.org/10.1002/hep.28549.

Ratzinger, G., et al., 2004. Mature human Langerhans cells derived from CD34+ hematopoietic progenitors stimulate greater cytolytic T lymphocyte activity in the absence of bioactive IL-12p70, by either single peptide presentation or cross-priming, than do dermal-interstitial or monocyte-derived dendritic cells. J. Immunol. 173, 2780–2791. https://doi.org/10.4049/jimmunol.173.4.2780.

Redman, B.G., et al., 2008. Phase Ib trial assessing autologous, tumor-pulsed dendritic cells as a vaccine administered with or without IL-2 in patients with metastatic melanoma. J. Immunother. 31, 591–598. https://doi.org/10.1097/CJI.0b013e31817fd90b.

Regnault, A., et al., 1999. Fcgamma receptor-mediated induction of dendritic cell maturation and major histocompatibility complex class I-restricted antigen presentation after immune complex internalization. J. Exp. Med. 189, 371–380. https://doi.org/10.1084/jem.189.2.371.

Ribas, A., et al., 2005. Antitumor activity in melanoma and anti-self responses in a phase I trial with the anti-cytotoxic T lymphocyte-associated antigen 4 monoclonal antibody CP-675,206. J. Clin. Oncol. 23, 8968–8977. https://doi.org/10.1200/JCO.2005.01.109.

Ribas, A., et al., 2009. Dendritic cell vaccination combined with CTLA4 blockade in patients with metastatic melanoma. Clin. Cancer Res. 15, 6267–6276. https://doi.org/10.1158/1078-0432.CCR-09-1254.

Richter, C., et al., 2013. Generation of inducible immortalized dendritic cells with proper immune function in vitro and in vivo. PLoS One 8, e62621. https://doi.org/10.1371/journal.pone.0062621.

Roberts, E.W., et al., 2016. Critical role for CD103(+)/CD141(+) dendritic cells bearing CCR7 for tumor antigen trafficking and priming of T cell immunity in melanoma. Cancer Cell 30, 324–336. https://doi.org/10.1016/j.ccell.2016.06.003.

Romano, E., et al., 2011. Peptide-loaded Langerhans cells, despite increased IL15 secretion and T-cell activation in vitro, elicit antitumor T-cell responses comparable to peptide-loaded monocyte-derived dendritic cells in vivo. Clin. Cancer Res. 17, 1984–1997. https://doi.org/10.1158/1078-0432.CCR-10-3421.

Sabado, R.L., Balan, S., Bhardwaj, N., 2017. Dendritic cell-based immunotherapy. Cell Res. 27, 74–95. https://doi.org/10.1038/cr.2016.157.

Saberian, C., et al., 2021. Randomized phase II trial of lymphodepletion plus adoptive cell transfer of tumor-infiltrating lymphocytes, with or without dendritic cell vaccination, in patients with metastatic melanoma. J. Immunother. Cancer 9. https://doi.org/10.1136/jitc-2021-002449.

Salazar, A.M., Erlich, R.B., Mark, A., Bhardwaj, N., Herberman, R.B., 2014. Therapeutic in situ autovaccination against solid cancers with intratumoral poly-ICLC: case report, hypothesis, and clinical trial. Cancer Immunol. Res. 2, 720–724. https://doi.org/10.1158/2326-6066.CIR-14-0024.

Sallusto, F., Cella, M., Danieli, C., Lanzavecchia, A., 1995. Dendritic cells use macropinocytosis and the mannose receptor to concentrate macromolecules in the major histocompatibility complex class II compartment: downregulation by cytokines and bacterial products. J. Exp. Med. 182, 389–400. https://doi.org/10.1084/jem.182.2.389.

Sanmamed, M.F., Chen, L., 2018. A paradigm shift in cancer immunotherapy: from enhancement to normalization. Cell 175, 313–326. https://doi.org/10.1016/j.cell.2018.09.035.

Savina, A., Amigorena, S., 2007. Phagocytosis and antigen presentation in dendritic cells. Immunol. Rev. 219, 143–156. https://doi.org/10.1111/j.1600-065X.2007.00552.x.

Schmitz, M., et al., 2002. Native human blood dendritic cells as potent effectors in antibody-dependent cellular cytotoxicity. Blood 100, 1502–1504.

Schmitz, M., et al., 2005. Tumoricidal potential of native blood dendritic cells: direct tumor cell killing and activation of NK cell-mediated cytotoxicity. J. Immunol. 174, 4127–4134. https://doi.org/10.4049/jimmunol.174.7.4127.

Schreibelt, G., et al., 2016. Effective clinical responses in metastatic melanoma patients after vaccination with primary myeloid dendritic cells. Clin. Cancer Res. 22, 2155–2166. https://doi.org/10.1158/1078-0432.CCR-15-2205.

Shen, Z., Reznikoff, G., Dranoff, G., Rock, K.L., 1997. Cloned dendritic cells can present exogenous antigens on both MHC class I and class II molecules. J. Immunol. 158, 2723–2730.

Sheng, K.C., Pietersz, G.A., Wright, M.D., Apostolopoulos, V., 2005. Dendritic cells: activation and maturation—applications for cancer immunotherapy. Curr. Med. Chem. 12, 1783–1800. https://doi.org/10.2174/0929867054367248.

Shimizu, K., Fields, R.C., Giedlin, M., Mule, J.J., 1999. Systemic administration of interleukin 2 enhances the therapeutic efficacy of dendritic cell-based tumor vaccines. Proc. Natl. Acad. Sci. U. S. A. 96, 2268–2273. https://doi.org/10.1073/pnas.96.5.2268.

Shimizu, K., Fields, R.C., Redman, B.G., Giedlin, M., Mule, J.J., 2000. Potentiation of immunologic responsiveness to dendritic cell-based tumor vaccines by recombinant interleukin-2. Cancer J. Sci. Am. 6 (Suppl. 1), S67–S75.

Sichien, D., Lambrecht, B.N., Guilliams, M., Scott, C.L., 2017. Development of conventional dendritic cells: from common bone marrow progenitors to multiple subsets in peripheral tissues. Mucosal Immunol. 10, 831–844. https://doi.org/10.1038/mi.2017.8.

Sioud, M., et al., 2013. Silencing of indoleamine 2,3-dioxygenase enhances dendritic cell immunogenicity and antitumour immunity in cancer patients. Int. J. Oncol. 43, 280–288. https://doi.org/10.3892/ijo.2013.1922.

Small, E.J., et al., 2000. Immunotherapy of hormone-refractory prostate cancer with antigen-loaded dendritic cells. J. Clin. Oncol. 18, 3894–3903. https://doi.org/10.1200/JCO.2000.18.23.3894.

Small, E.J., et al., 2006. Placebo-controlled phase III trial of immunologic therapy with sipuleucel-T (APC8015) in patients with metastatic, asymptomatic hormone refractory prostate cancer. J. Clin. Oncol. 24, 3089–3094. https://doi.org/10.1200/JCO.2005.04.5252.

Sprooten, J., et al., 2019. Trial watch: dendritic cell vaccination for cancer immunotherapy. Oncoimmunology 8, e1638212. https://doi.org/10.1080/2162402X.2019.1638212.

Srivastava, R.M., Varalakshmi, C., Khar, A., 2007. Cross-linking a mAb to NKR-P2/NKG2D on dendritic cells induces their activation and maturation leading to enhanced anti-tumor immune response. Int. Immunol. 19, 591–607. https://doi.org/10.1093/intimm/dxm024.

Steinman, R.M., Cohn, Z.A., 1973. Identification of a novel cell type in peripheral lymphoid organs of mice. I. Morphology, quantitation, tissue distribution. J. Exp. Med. 137, 1142–1162. https://doi.org/10.1084/jem.137.5.1142.

Steinman, R.M., Hemmi, H., 2006. Dendritic cells: translating innate to adaptive immunity. Curr. Top. Microbiol. Immunol. 311, 17–58. https://doi.org/10.1007/3-540-32636-7_2.

Tay, R.E., Richardson, E.K., Toh, H.C., 2021. Revisiting the role of CD4(+) T cells in cancer immunotherapy-new insights into old paradigms. Cancer Gene Ther. 28, 5–17. https://doi.org/10.1038/s41417-020-0183-x.

Tel, J., et al., 2013. Natural human plasmacytoid dendritic cells induce antigen-specific T-cell responses in melanoma patients. Cancer Res. 73, 1063–1075. https://doi.org/10.1158/0008-5472.CAN-12-2583.

Terhune, J., Berk, E., Czerniecki, B.J., 2013. Dendritic cell-induced Th1 and Th17 cell differentiation for cancer therapy. Vaccines (Basel) 1, 527–549. https://doi.org/10.3390/vaccines1040527.

Thery, C., Amigorena, S., 2001. The cell biology of antigen presentation in dendritic cells. Curr. Opin. Immunol. 13, 45–51. https://doi.org/10.1016/s0952-7915(00)00180-1.

Tkach, M., et al., 2017. Qualitative differences in T-cell activation by dendritic cell-derived extracellular vesicle subtypes. EMBO J. 36, 3012–3028. https://doi.org/10.15252/embj.201696003.

Trakatelli, M., et al., 2006. A new dendritic cell vaccine generated with interleukin-3 and interferon-beta induces CD8+ T cell responses against NA17-A2 tumor peptide in melanoma patients. Cancer Immunol. Immunother. 55, 469–474. https://doi.org/10.1007/s00262-005-0056-z.

Tran Janco, J.M., Lamichhane, P., Karyampudi, L., Knutson, K.L., 2015. Tumor-infiltrating dendritic cells in cancer pathogenesis. J. Immunol. 194, 2985–2991. https://doi.org/10.4049/jimmunol.1403134.

Treilleux, I., et al., 2004. Dendritic cell infiltration and prognosis of early stage breast cancer. Clin. Cancer Res. 10, 7466–7474. https://doi.org/10.1158/1078-0432.CCR-04-0684.

Trepiakas, R., et al., 2010. Vaccination with autologous dendritic cells pulsed with multiple tumor antigens for treatment of patients with malignant melanoma: results from a phase I/II trial. Cytotherapy 12, 721–734. https://doi.org/10.3109/14653241003774045.

Vacchelli, E., et al., 2015. Chemotherapy-induced antitumor immunity requires formyl peptide receptor 1. Science 350, 972–978. https://doi.org/10.1126/science.aad0779.

Valmori, D., Ayyoub, M., 2004. Using modified antigenic sequences to develop cancer vaccines: are we losing the focus? PLoS Med. 1, e26. https://doi.org/10.1371/journal.pmed.0010026.

van Beek, J.J., Wimmers, F., Hato, S.V., de Vries, I.J., Skold, A.E., 2014. Dendritic cell cross talk with innate and innate-like effector cells in antitumor immunity: implications for DC vaccination. Crit. Rev. Immunol. 34, 517–536. https://doi.org/10.1615/critrevimmunol.2014012204.

van de Laar, L., Lambrecht, B.N., 2014. How to generate large numbers of CD103+ dendritic cells. Blood 124, 3036–3038. https://doi.org/10.1182/blood-2014-08-595298.

van de Loosdrecht, A.A., et al., 2018. A novel allogeneic off-the-shelf dendritic cell vaccine for post-remission treatment of elderly patients with acute myeloid leukemia. Cancer Immunol. Immunother. 67, 1505–1518. https://doi.org/10.1007/s00262-018-2198-9.

van der Waart, A.B., et al., 2015. siRNA silencing of PD-1 ligands on dendritic cell vaccines boosts the expansion of minor histocompatibility antigen-specific CD8(+) T cells in NOD/SCID/IL2Rg(null) mice. Cancer Immunol. Immunother. 64, 645–654. https://doi.org/10.1007/s00262-015-1668-6.

Van Hoecke, L., et al., 2021. mRNA in cancer immunotherapy: beyond a source of antigen. Mol. Cancer 20, 48. https://doi.org/10.1186/s12943-021-01329-3.

Van Lint, S., et al., 2016. Intratumoral delivery of TriMix mRNA results in T-cell activation by cross-presenting dendritic cells. Cancer Immunol. Res. 4, 146–156. https://doi.org/10.1158/2326-6066.CIR-15-0163.

Vik-Mo, E.O., et al., 2013. Therapeutic vaccination against autologous cancer stem cells with mRNA-transfected dendritic cells in patients with glioblastoma. Cancer Immunol. Immunother. 62, 1499–1509. https://doi.org/10.1007/s00262-013-1453-3.

Vo, M.C., et al., 2017. Combination therapy with dendritic cells and lenalidomide is an effective approach to enhance antitumor immunity in a mouse colon cancer model. Oncotarget 8, 27252–27262. https://doi.org/10.18632/oncotarget.15917.

Wang, L., et al., 2020. Naringenin enhances the antitumor effect of therapeutic vaccines by promoting antigen cross-presentation. J. Immunol. 204, 622–631. https://doi.org/10.4049/jimmunol.1900278.

Wculek, S.K., et al., 2019. Effective cancer immunotherapy by natural mouse conventional type-1 dendritic cells bearing dead tumor antigen. J. Immunother. Cancer 7, 100. https://doi.org/10.1186/s40425-019-0565-5.

Wickström, S.L., et al., 2018. Adoptive T cell transfer combined with DC vaccination in patients with metastatic melanoma. Cancer Res. 78, CT032. https://doi.org/10.1158/1538-7445.AM2018-CT032.

Wilgenhof, S., et al., 2016. Phase II study of autologous monocyte-derived mRNA electroporated dendritic cells (TriMixDC-MEL) plus ipilimumab in patients with pretreated advanced melanoma. J. Clin. Oncol. 34, 1330–1338. https://doi.org/10.1200/JCO.2015.63.4121.

Wolkers, M.C., Stoetter, G., Vyth-Dreese, F.A., Schumacher, T.N., 2001. Redundancy of direct priming and cross-priming in tumor-specific CD8+ T cell responses. J. Immunol. 167, 3577–3584. https://doi.org/10.4049/jimmunol.167.7.3577.

Xin, Y.J., Hubbard-Lucey, V.M., Tang, J., 2019. The global pipeline of cell therapies for cancer. Nat. Rev. Drug Discov. 18, 821–822. https://doi.org/10.1038/d41573-019-00090-z.

Xu, M.M., et al., 2017. Dendritic cells but not macrophages sense tumor mitochondrial DNA for cross-priming through signal regulatory protein alpha signaling. Immunity 47, 363–373 e365. https://doi.org/10.1016/j.immuni.2017.07.016.

Xu, Z., Zeng, S., Gong, Z., Yan, Y., 2020. Exosome-based immunotherapy: a promising approach for cancer treatment. Mol. Cancer 19, 160. https://doi.org/10.1186/s12943-020-01278-3.

Yang, S.C., et al., 2004. Intratumoral administration of dendritic cells overexpressing CCL21 generates systemic antitumor responses and confers tumor immunity. Clin. Cancer Res. 10, 2891–2901. https://doi.org/10.1158/1078-0432.ccr-03-0380.

Yu, Y., et al., 2001. Enhancement of human cord blood CD34+ cell-derived NK cell cytotoxicity by dendritic cells. J. Immunol. 166, 1590–1600. https://doi.org/10.4049/jimmunol.166.3.1590.

Zhao, L., et al., 2021. A genotype-phenotype screening system using conditionally immortalized immature dendritic cells. STAR Protoc. 2, 100732. https://doi.org/10.1016/j.xpro.2021.100732.

CHAPTER THREE

Killers on the loose: Immunotherapeutic strategies to improve NK cell-based therapy for cancer treatment

Cordelia Dunai[a,†], Erik Ames[b,†], Maria C. Ochoa[c,d,e], Myriam Fernandez-Sendin[c,d], Ignacio Melero[c,d,e,f], Federico Simonetta[g,h], Jeanette Baker[i], and Maite Alvarez[c,d,e,*]

[a]Department of Clinical Infection, Microbiology and Immunology, University of Liverpool, Liverpool, United Kingdom
[b]Department of Pathology, Stanford University, Stanford, CA, United States
[c]Program for Immunology and Immunotherapy, CIMA, Universidad de Navarra, Pamplona, Spain
[d]Navarra Institute for Health Research (IdiSNA), Pamplona, Spain
[e]Centro de Investigación Biomédica en Red de Cáncer (CIBERONC), Madrid, Spain
[f]Department of Immunology and Immunotherapy, Clínica Universidad de Navarra, Pamplona, Spain
[g]Division of Hematology, Department of Oncology, Geneva University Hospitals, Geneva, Switzerland
[h]Translational Research Centre in Onco-Haematology, Faculty of Medicine, Department of Pathology and Immunology, University of Geneva, Geneva, Switzerland
[i]Blood and Marrow Transplantation, Stanford University School of Medicine, Stanford, CA, United States
*Corresponding author: e-mail address: malvarezr@unav.es

Contents

1. Introduction 66
2. Immunomodulatory strategies 69
   2.1 Modulation of chemokines and cytokines to increase NK cell activation 69
   2.2 mRNA technology to modulate NK cell activation 75
   2.3 Modulation of immunosuppression to prevent NK cell inhibition 76
3. NK-based cellular therapy 83
   3.1 NK cells in hematopoietic stem cell transplantation 83
   3.2 Chimeric antigen receptor (CAR) and engineered NK cells 84
4. NK immune cell engagers 90
5. Oncolytic virotherapy 93
6. Conclusions 97
Acknowledgments 99
Conflict of interest 99
References 99

[†] CD and EA share first co-authorship.

Abstract

Natural killer (NK) cells are innate lymphocytes that control tumor progression by not only directly killing cancer cells, but also by regulating other immune cells, helping to orchestrate a coordinated anti-tumor response. However, despite the tremendous potential that this cell type has, the clinical results obtained from diverse NK cell-based immunotherapeutic strategies have been, until recent years, rather modest. The intrinsic regulatory mechanisms that are involved in the control of their activation as well as the multiple mechanisms that tumor cells have developed to escape NK cell-mediated cytotoxicity likely account for the unsatisfactory clinical outcomes. The current approaches to improve long-term NK cell function are centered on modulating different molecules involved in both the activation and inhibition of NK cells, and the latest data seems to advocate for combining strategies that target multiple aspects of NK cell regulation. In this review, we summarize the different strategies (such as engineered NK cells, CAR-NK, NK cell immune engagers) that are currently being used to take advantage of this potent and complex immune cell.

1. Introduction

Natural killer (NK) cells and their spontaneous anti-tumor cytotoxic activity were first described by two independent groups in 1975 (Herberman et al., 1975; Kiessling et al., 1975). Their descriptive name was coined by Eva Klein and although they have more recently been classified within the innate lymphoid cell group (ILC, type 1), they remain of great interest for therapeutic applications, especially for cancer (Spits et al., 2016). Even with the advent of advanced adoptive T cell treatment, i.e., chimeric antigen receptor (CAR) T cells, the challenge of immune evasion by tumor cells—including by antigen loss and/or downregulation of MHC—remains. In addition to NK cells' ability to target cells lacking MHC I expression, advantages of NK cells for cancer treatment include their cytotoxic effects and their protective role against graft-*versus*-host-disease (Ruggeri et al., 2002). As it will be discussed in this review, there is ongoing NK-cell research to overcome the issues of optimum activation prior to and during adoptive transfer, maximizing tumor infiltration, persistence, and tumor killing.

NK cells are lymphoid lineage cells that are part of the innate immune response. They make up 5–10% of peripheral blood mononuclear cells and have a significant tissue-resident population that plays a role in immune surveillance. In humans, they are also found in lymph nodes with the $CD56^{bright}CD16^{low}$ subpopulation dominating there, versus $CD56^{dim}CD16^{brigh}$ cells which predominate in circulation (Dogra et al., 2020). NK cells ontogeny occurs in the bone marrow and they do not require priming phase to take cytotoxic and proinflammatory action and are thought to be short-lived as compared to T cells. Studies have shown that their lifespan is approximately 14 days, however, tissue-resident populations have been found to be maintained for more than a decade in transplant recipients (Lahoz-Beneytez et al., 2017). Although they have roles in modulating other cells by cytokine production (DC, T cells), their primary role is their ability to kill malignant and virally infected cells. They can do this by four different mechanisms: Fas:FasL interactions, TRAIL and death receptor binding, perforin, granzyme release, and antibody-dependent cell mediated cytotoxicity (ADCC). The latter form of cytotoxicity has been critical for the clinical success of tumor-targeting monoclonal antibody (mAb) therapies including rituximab, cetuximab, trastuzumab or daratumumab, that target CD20, EGFR, HER2 and CD38 on lymphoma, colon cancer, breast cancer and multiple myeloma respectively (Gauthier et al., 2021). ADCC might be the most important function of NK cells and is mediated by the recognition of antibody coated cells by CD16A (FcγRIIIA).

With this cell-destructive power, it is important for safe regulation to ensure self-tolerance and therefore NK cells present intrinsic regulatory mechanisms to ensure that they are not improperly triggered. First, during ontogeny, NK cells undergo a process named "licensing," "arming" or "education." Epigenetic mechanisms lead to the clonal expression of inhibitory receptors in a stochastic fashion (Anfossi et al., 2006; Kim et al., 2005; Raulet, 2006). This phenomenon leads to the generation of two distinct NK cell subsets: licensed NK cells, which express inhibitory receptors against self-MHC class I alleles (MHC-I) and can be shutdown later in presence of normal cells, and unlicensed NK cells which do not express at least one inhibitory receptor for MHC-I. Licensed NK cells are better equipped to respond against transformed cells (Anfossi et al., 2006; Kim et al., 2005; Raulet, 2006). However, in an inflammatory environment, the killing capacities of both subsets are indistinct and can equally exert their effector functions, which have proven advantageous for allogeneic bone marrow transplants (Alvarez et al., 2020a; Barao et al., 2011; Sun et al., 2012; Yu et al., 2009). Second, the regulation of mature NK cell is mediated by balancing the signals coming from activating receptors (NKG2D, NKG2C, Killer Immunoglobulin-like Receptor [KIR], NKp46, NKp30, 2B4), which recognize stress signals expressed on transformed cells; and inhibitory receptors (NKG2A, KIRs, CD96, TIGIT, TIM3), which recognize MHC class I-derived molecules on normal cells, along signals from immune checkpoint inhibitors (like PD-1/PD-L1, CTLA4) (Hodgins et al., 2019; Myers and Miller, 2021; Pierini et al., 2016; Simonetta et al., 2017) and stimulating receptors (such as CD137, OX40, TNFR) (Mancusi et al., 2019). The net signaling from these receptors, along with the surrounding cytokine environment, will determine the functional fate of NK cells, which are by default unresponsive (Fig. 1). Once activated, these

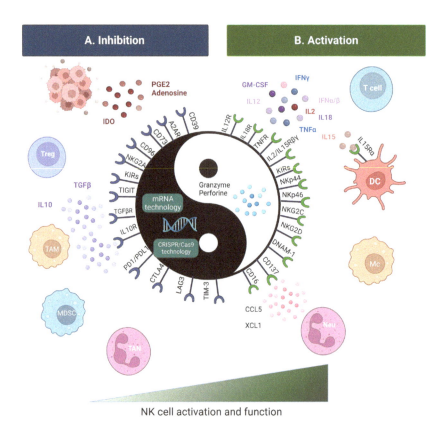

Fig. 1 Regulatory mechanisms for the modulation of NK cell activation and function that can be therapeutically targeted. (A) Immunosuppressor cells (tumors, Tregs, myeloid-derived suppressor cells-MDSC, tumor-associated macrophages-TAMs, and tumor-associated neutrophils-TANs), molecules (IDO, PGE2, adenosine, TGFβ and IL-10, CD39, CD73), and inhibitory receptors (inhibitory KIRs, NKG2A, TIGIT, CD96, A2AR, PD-1/PD-L1, CTLA-4, LAG3 and TIM-3) involved in the inhibition of NK cells. (B) Activating immune cells (T cells, DC, macrophages and neutrophils), cytokines (GM-CSF, IL-2, IL-12, IL-15, IL-18, type I IFN, TNFα), chemokines (CCL5, XCL1) and receptors (NKp44, NKp46, NKG2C, NKG2D, CD16, CD137, DNAM-1) involved in the activation and acquisition of function of NK cells. *Figure created with BioRender.com.*

same mechanisms are used to prevent exacerbated responses as the repertoire of inhibitory and activating receptors tends to change towards a more easily suppressive subset with marked upregulation of inhibitory receptors (Alvarez et al., 2019). In settings of sustained chronic stimulation, NK cells can undergo exhaustion, anergy and/or senescence that renders then dysfunctional and incapable of responding to subsequent stimuli (Alvarez et al., 2019; Felices et al., 2018; Judge et al., 2020a).

Immune suppression mediated by tumor cells represents one of the major challenges to immunotherapy success and it is likely the main factor for the clinical failure of many NK cell-based immunotherapy approaches. Tumors can mimic most of the regulatory mechanisms used by NK cells and thereby can dampen NK function by the recruitment of immunosuppressive cells (regulatory T cells [Tregs], myeloid-derived suppressive cells [MDSC], tumor associated macrophages [TAMs] or tumor-associated neutrophils [TANs]) and the release of suppressive molecules such as IL-10 and TGF-beta, indoleamine 2, 3-dioxygenase (IDO) and prostaglandin E2 (PGE2), as well as by triggering immune checkpoint inhibitors and promote exhaustion among many others (Chauhan et al., 2020; Li et al., 2012; Myers and Miller, 2021; Song et al., 2021). A therapeutic strategy to avoid these issues is to make NK cells functionally active and more resistant to suppression. In this review, we will revisit some of the immunotherapeutic NK cell-based strategies that are currently under research to modulate NK cell function in order to achieve strong and durable anti-tumor responses (Fig. 1).

2. Immunomodulatory strategies

2.1 Modulation of chemokines and cytokines to increase NK cell activation

2.1.1 Chemokines

As with other cellular therapies, a major challenge to successful tumor therapy is guiding the adoptively transferred cells (most often infused into blood) to traffic and infiltrate the tumor. Research on the chemokines and receptors necessary for optimum NK cell homing are ongoing. Oncolytic viruses, which will be discussed later, have the promising property of causing tumor lysis and consequent release of chemotactic factors. Chemokines including CCL2, CXCL9/10/11, and CX3CL1 provide signals for NK cell migration into tissue. However, the challenge remains to increase the expression of these cytokines within the tumor to attract NK cells. Work is ongoing to engineer NK cells to express chemokine receptors that would be beneficial for tumor infiltration. The cytokine stimulation regimens discussed below activate NK cells and can increase their chemokine receptor expression (Yao and Matosevic, 2021). However, given the role of chemokines in the recruitment of immunosuppressor cells, inhibition of chemokines might reduce tumor infiltration of these cells and allow the presence of effector cells. According to this notion, inhibition of CXCR1/2 by SX-682 resulted in a reduction of MDSC in head and neck squamous cell carcinoma mouse tumor model while enhanced the tumor infiltration, activation, and therapeutic efficacy of adoptively transferred NK cells (Greene et al., 2020).

There is burgeoning field dedicated to optimizing activation of NK cells prior to and during adoptive transfer. NK cells are particularly dependent on cytokine signaling for survival and function. The cytokines that signal through common gamma chain (including IL-2, IL-4, IL-7, IL-9, IL-15, IL-21) provide a survival and activation signal to NK cells, triggering the JAK/STAT pathway, and allowing for proliferation and effector functions. Although the short lifespan of NK cells is a challenge, this might be advantageous from a safety perspective. Since NK cells are not restricted to MHC antigen presentation, they are also attractive for "off-the-shelf" therapeutic application as will be discuss later on.

2.1.2 IL-2

IL-2 is a cytokine that signals through the high affinity trimeric receptor formed by the IL-2Rα, IL-2Rβ and the common γ chain, or through the low affinity dimer formed by IL-2Rβ and the common γ chain. IL-2 is an activation, maturation and proliferation factor of lymphocytes and induces the activation and expansion of NK cells when directly administered in preclinical (Alvarez et al., 2014, 2020b; Hallett et al., 2008) and clinical models (Miller et al., 2005). Unfortunately, in order to achieve biological significance, the dosage needed is very high and has resulted in important treatment-related adverse events, including vascular leak syndrome, heart failure and liver toxicity, despite achieving around 10% complete responses in some studies (Payne et al., 2014); and thus, its implementation in the clinic as a monotherapy has been limited. To avoid those complications, some strategies are being explored, such as local tumor injection, low-dose regimens, administration in nanoparticles or modifications in the molecule to avoid endothelial binding and targeting to the tumor (Boieri et al., 2017; Gillies et al., 2011; Kim et al., 2022; Miller et al., 1997; Nguyen et al., 2019).

Since the pioneer clinical trials of lymphokine-activated killer (LAK) cell adoptive therapy performed by Rosenberg et al., IL-2 has been typically administered together with the LAK, T and NK cells to increase the efficiency and persistence of the cells (Rosenberg et al., 1986). However, IL-2 induces the activation of Tregs (via the constitutively expressed IL-2Rα,) which would be detrimental in the cancer setting (Hallett et al., 2008). Alternative molecules that carry a modified IL-2 are being developed to decrease this immunosuppressor effect. Recently, Sahin et al. designed a fusion protein consisting of IL-2 linked to a part of IL-2Rα which abolishes the binding of IL-2 to IL-2Rα, while also increasing the *in vivo* half-life of IL-2 (Sahin et al., 2020). Following a similar concept, a fusion protein formed by a bispecific IL-2v immunocytokine that lacks binding to IL-2Rα while retains IL-2R$\beta\gamma$ binding, and fused to the fibrobroblast activation protein-α named RO687428 is being explored in several clinical trials (NCT03875079, NCT02627274). This crafted IL-2 antibody induced a potent expansion of CD8 T and NK cells, which were responsible for strong anti-tumor responses in preclinical tumor models (Sahin et al., 2020). Alternatively, an engineered semi-synthetic organism-derived IL-2 variant (THOR-707), currently under evaluation on two clinical trials (NCT04009681, NCT05179603), has showed a longer half-life and lower IL-2Rα engagement (Ptacin et al., 2021). Another reoccurring challenge for investigating stimulation strategies with IL-2 is NK cell exhaustion and the resulting impaired functional activity observed after prolonged stimulation (Alvarez et al., 2019, 2020c), suggesting that the use of IL-2 on its own might not be sufficient to achieve sustained NK cell responses and overcome dysfunction.

2.1.3 IL-15

IL-15, similar to IL-2, signals through the common γ chain receptor and leads to NK cell activation and proliferation. IL-15 is trans-presented with IL-15Rα on dendritic cells and stromal cells to the IL-2Rβ and γ common chain heterodimeric receptor on T and NK cells. In physiological conditions, IL-15 is presented by dendritic cells to neighboring NK and T cells adsorbed onto IL-15Rα. This phenomenon is termed trans-presentation of IL-15. As with IL-2, prolonged IL-15 stimulation also causes NK cell exhaustion (Alvarez et al., 2019; Elpek et al., 2010; Uppendahl et al., 2019). However, an advantage of IL-15 over IL-2 is that IL-15 avoids the direct activation of Tregs, making it the cytokine of choice frequently used for both the *in vitro* and *in vivo* NK cell expansion (Alvarez et al., 2020d). Some clinical trials administering IL-15 have been developed in diverse types of cancers, but the overall anti-tumor effect has been very limited at the tolerated doses (Waldmann et al., 2020). Diverse regimens and administration routes have been tested; with subcutaneous injection more tolerated than intravenous bolus. Recently, Dubois et al. have reported that 5 days infusion of 5 µg/kg/day is safer than other tested doses and obtained similar pharmacologic effects (NCT01572493, NCT03759184, NCT03905135, NCT04185220 and NCT02689453) (Dubois et al., 2021). However, recombinant proteins that fuse IL-15 to IL-15Rα have proven more bioactive and stable. In addition, several new compounds mimicking the presentation with chimeric proteins are under development (Bessard et al., 2009; Chertova et al., 2013; Guo et al., 2021a; Mortier et al., 2006; Ochoa et al., 2013, 2018; Schmohl et al., 2016a). The super-agonist IL-15 N-803 (formerly ALT-803) (Furuya et al., 2019) is a fusion protein encompassing IL-15, the IL-15Rα and a modified human IgG that does not induce ADCC. *In vitro* and preclinical studies have shown that this molecule exerts an anti-tumor effect through T and NK cells. Moreover, N-803 was capable of reverting the immunosuppressive effect of TGFβ on NK cells (Fujii et al., 2018). In a clinical trial performed in patients with solid cancers, N-803 has shown to be safe and well tolerated, and causes an increase in circulating NK cells (NCT01727076) (Margolin et al., 2018).

The Miller group at the University of Minnesota have, in large part, pioneered NK cell transfers with concomitant cytokine stimulation (Miller et al., 2005). Two key clinical studies combined lymphodepleting chemotherapy and haploidentical NK cells with subcutaneous or intravenous administration of IL-15 in patients with advanced acute myeloid leukemia, achieving 40% remission with acceptable toxicity issues (Cooley et al., 2019). However, another study that compared side-by-side IL-2 and N-803 combined with lymphodepleting chemotherapy and haploidentical NK cells, also in relapsed/refractory acute myeloid leukemia patients, failed to show a benefit from the IL-15 cohort as 14% of patients that received IL-15 achieved complete response *vs* the 28% from the IL-2 cohort (Berrien-Elliott et al., 2021). Circulating NK cells of patients from both groups where similar at day 21 after infusion, but the numbers decreased faster in the N-803 group (Berrien-Elliott et al., 2021). Nevertheless, IL-15 and their modified molecules still remain a promising tool for NK-based cancer treatment, but NK cells can also become exhausted after sustained IL-15 stimulation, (Alvarez et al., 2019, 2020c; Elpek et al., 2010; Felices et al., 2018), therefore, optimized dose, timing, and combinatorial treatment might be required.

2.1.4 IL-12

IL-12 was originally named as NK cell stimulatory factor and it induces IFNγ as well as perforin and granzyme B production, as well as enhancing NK cell cytotoxicity (Aste-Amezaga et al., 1994). Clinical development of IL-12 started in patients with renal cell carcinoma and melanoma with certain indications of clinical activity. However, a phase II trial in patients with renal cancer resulted in severe adverse effects within the first 5 days of treatment. Twelve out of 17 enrolled patients required hospitalization with intensive care and two patients died (Conlon et al., 2019; Leonard et al., 1997). After that, clinical trials administering safer regimens of IL-12 have been developed in patients with renal cell carcinoma and peritoneal carcinomatosis with modest effects (Lenzi et al., 2007; Motzer et al., 1998). The NHS-IL12 immunocytokine that targets IL-12 to the tumor has been also tested in patients with metastatic solid tumors and the treatment was well-tolerated and enhanced immune-related activity was shown, even though no clinical responses were observed (Strauss et al., 2019).

Because NK cells can exert their cytotoxicity better against antibody-coated tumor cells (ADCC), it is of great interest to study additive effects of combining IL-12 treatment with antibody-based therapies that target tumors (Alderson and Sondel, 2011; Parihar et al., 2002). A clinical trial combining cetuximab and IL-12 in patients with unresectable primary or recurrent head and neck squamous cell carcinoma has been performed with, unfortunately, no evidence of objective clinical responses (McMichael et al., 2019). Despite this, 48% of patients still exhibited prolonged progression-free survival. In another trial that tested paclitaxel and trastuzumab in combination with IL-12 in patients with HER2/neu-expressing malignancies, 50% of the patients showed clinical benefits and higher levels of serum IFN-γ were found in those patients (Bekaii-Saab et al., 2009). At present, more combinations and safer modified IL-12 versions are being developed with different strategies such as INO-9012, a DNA-based expression plasmid DNA encoding IL-12 (NCT03502785) and mRNAs encoding IL-12 (Cirella et al., 2020).

2.1.5 Cytokine combinations

It has been shown that the combination of IL-12/15/18 potently activates NK cells. The use of these three cytokines mimics a potent innate immune signal and triggers two main signaling pathways STAT3, 4, 5 and MyD88 leading to NF-κB activation and downstream activation and effector functions. Combinatorial pre-activation generates "cytokine-induced memory-like (CIML)" NK cells, which have demonstrated increased IFNγ production, proliferation, and anti-tumor effects (Boieri et al., 2017; Uppendahl et al., 2019). Some clinical trials have been performed transferring these CIML-NK cells into patients with acute myeloid leukemia, achieving four out of nine complete remissions (NCT01898793) (Romee et al., 2016).

2.1.6 TNFR superfamily members

Many studies have shown the role of the members of the TNFR superfamily (CD137 [41BB], OX40, TNFR1/TNFR2) in the regulation of NK cell activation as well T cells and DCs, who also express these stimulatory receptors (Guo et al., 2021b; Mancusi et al., 2019; Sanmamed et al., 2015). These are currently being explored in many clinical trials as a form of agonistic mAbs, NK or T cell immune engagers, or added to CARs in order to contribute to the activation and expansion of these effector cells (Guo et al., 2021b; Mancusi et al., 2019; Myers and Miller, 2021; Sanmamed et al., 2015).

Triggering CD137 to expand NK cells is a common practice for *ex vivo* expansion of NK cells in clinical applications, which is the case for two clinical trials (NCT03841110 and NCT04106167) that use K562 feeder cells modified to express CD137 ligand (41BBL) and IL-21. A phase two clinical trial using haploidentical NK cells activated and expanded with membrane bound IL-15-41BBL K562 is also ongoing ((NCT02763475) (Gomez Garcia et al., 2021). Results from preclinical and clinical studies have also suggested promise for anti-CD137 agonistic antibodies due to their ability to enhance NK cell cytotoxicity and ADCC (Chu et al., 2019; Masu et al., 2018; Melero et al., 1997; Wilcox et al., 2002). This feature seems to be particularly beneficial when anti-CD137 is combined with tumor-targeting antibody-based therapies such as cetuximab in colorectal and head and neck cancer or trastuzumab in breast cancer (Srivastava et al., 2017). In the same line, *in vitro* data has shown stronger ADCC against multiple myeloma cell lines from NK cells stimulated with daratumumab (anti-CD38) acting on CD16 and urelumab (agonist anti-CD137) (Ochoa et al., 2019).

On the other hand, a recent study has suggested that CD137 co-stimulation of NK cell cultures can also counteract the negative influence of TGFβ by preserving the expression of NKG2D and its response to IL-2 in terms of proliferation and cytokine production (Cabo et al., 2021). However, the anti-tumor benefits of CD137 agonistic antibodies might be partially attributed to NK cells (Melero et al., 1998) and still remains a topic of current investigation in human trials with 4-1BB agonists, as anti-CD137 treatment acts on T cells and perhaps on DCs as well (Otano et al., 2021). It is likely that the three immune cell types collaborate to mediate the anti-tumor response (Srivastava et al., 2017). Similar to CD137, preclinical studies have shown that an OX40 mAb agonist can synergize with anti-CD20 therapy to enhance NK cell-mediated ADCC (Turaj et al., 2018). Triggering both OX40 and CD137 has also been explored (Guo et al., 2021b).

2.2 mRNA technology to modulate NK cell activation

Recent development in the field of messenger RNA (mRNA)-based therapies afford new approaches that can improve the safety and efficacy of immune drugs and prevent severe toxicities associated with systemic therapeutic strategies (Di Trani et al., 2022; Melero et al., 2021). mRNA serves as a tool for the expression of proteins by introducing exogenous molecules into target cells. *In vitro* transcribed mRNA structure is a single single-stranded (ss) composed of a Cap structure followed by a 5′ untranslated region (UTR), the open reading frame (ORF), 3′ UTR and the poly(A) tail (Beck et al., 2021; Di Trani et al., 2022; Pastor et al., 2018; Sahin et al., 2014). mRNA-based therapies are more attractive than other viral-based therapies because it produces a peak expression of the encoding protein shortly after administration (hours) and can be explored as an alternative to oncolytic therapy (Liu et al., 2016). Moreover, mRNA is degraded by normal cellular process, and it has a cytoplasmic delivery. Therefore, its short half-life results in transient and controlled expression of the encoding protein, resulting in the reduction of toxicities associated with treatment (Foster et al., 2019; Pardi et al., 2018; Sahin et al., 2014; Schlake et al., 2019). mRNA strategy is thereby being investigated as a tool for the transient modulation of immune cells.

Several studies have shown that mRNA electroporation improves NK cell homing. For example, Levy et al. demonstrated that after transfection with mRNA encoding CD34, nearly 100% of NK cell expressed this marker on the cell surface for up to 5 days without altering expression of activating and inhibitory NK cell receptors (Levy et al., 2016). Based on these findings, Carlsten and colleges electroporated primary NK cells with CCR7 mRNA. More than 95% of NK cells showed significant enhanced migration capacity and cytotoxicity activity (Carlsten et al., 2016). One more example of this approach is the enhanced bone marrow homing of NK cells after mRNA transfection with CXCR4. It is known that reduced CXCR4 expression is associated with impaired NK cell migration. To enhance the migratory capacity of NK cells, the authors genetically engineered *ex vivo* expanded NK cells to express CXCR4, resulting in increased cell surface expression of this chemokine receptor without altering the cell phenotype, cytotoxic function, or viability (Levy et al., 2019).

Cytokine-based immunotherapy can be also improved by mRNA strategies. In this regard, intratumoral injections of IL-12 mRNA augmented the IFNγ production, increasing consequently the expression of T- and NK-cell activation genes, that result in robust T cell- and NK cell-dependent anti-tumor responses and tumor regression (Di Trani et al., 2022; Hewitt et al., 2020). Similarly, a combination of mRNAs encoding IL-12 single chain (IL-12 sc), IL-15 fused to IL-15Rα, IFNα and GM-CSF is currently being tested in a clinical trial (NCT03871348). This mixture of mRNA, termed SAR440100 (BNT131), mediates successful antitumor immunity (Di Trani et al., 2022; Hewitt et al., 2020). According to these studies, mRNA technology immunomodulating NK cells provides a new route to modulate NK cell function and therefore improve the outcome of cancer immunotherapy using NK cells.

2.3 Modulation of immunosuppression to prevent NK cell inhibition

For an NK cell to become functional, it must scan the surrounding cells in order to recognize activating signals, which might tilt the scale towards activation. However, likely more important than the activating signals are the existence of inhibitory signals, which are critical in providing immunological self-tolerance and a negative-feedback mechanism to counteract the stimulatory signals and prevent an inadequate activation that can result in detrimental effects (Joncker et al., 2009; Myers and Miller, 2021). In order to scape NK cell recognition and activation, tumor cells have developed a variety of mechanisms that generally involve the creation of a strongly immunosuppressive environment. These mechanisms include recruiting immunosuppressor cells (Tregs, myeloid-derived suppressor cells, tumor-associated macrophages, tumor-associated neutrophils, fibroblast); that along with tumor cells can release immunosuppressor molecules known to inhibit NK cells (TGFβ, IL-10, IDO, PGE2); to upregulate NK cell inhibitory receptors (KIRs, NKG2A, TIGIT, CD96) and/or their corresponding major histocompatibility complex (MHC) (Chauhan et al., 2020; Li et al., 2012; Myers and Miller, 2021; Song et al., 2021). Other inhibitory immune checkpoints that have been described to influence NK cell activation outcomes and can be used by tumor cells to escape NK cell-mediated cytotoxicity include PD-1/PD-L1, CTLA-4, TIM-3 and LAG-3, whose expression has been highly correlated with NK cell dysfunction (Alvarez et al., 2020c; Bi and Tian, 2019; Chauhan et al., 2020; Dunai and Murphy, 2018; Hsu et al., 2018; Judge et al., 2020a; Srivastava et al., 2017; Vari et al., 2018). As the tumor microenvironment (TME) highly influences the fate on NK cells in solid tumors, we also have to consider hypoxia as an important factor for NK cell dysfunction (Melaiu et al., 2019). Many studies have shown that the expression of molecules involved in this process (IDO, PGE2, CD39, CD37, adenosine receptors) negatively impact NK cell activation and cytotoxicity (Chambers and Matosevic, 2019; Melaiu et al., 2019).

Taking into consideration the tremendous importance of NK cell inhibition for tumor progression, it is not surprising that many translational efforts have been concentrated in the development of immunotherapeutic strategies that attempt to modulate and prevent it (Bi and Tian, 2019; Chauhan et al., 2020; Sanchez-Correa et al., 2019; Zhang and Liu, 2020), as it is now known that robust NK activation is hardly enough to sustain NK cell functionality over time (Alvarez et al., 2019; Felices et al., 2018).

2.3.1 TGFβ

TGFβ is a pleiotropic group of three cytokines produced by many immune cells and non-immune cells as well as the tumor cells themselves that is known to functionally inhibit NK cells (Barao et al., 2006; Hallett et al., 2008; Marcoe et al., 2012; Tauriello et al., 2022). Among other activities, TGFβ impairs NK cell abilities to secrete cytokines and degranulate, their metabolism and mTOR signaling. Because of its importance, TGFβ neutralization has been extensively explored for clinical applications with many strategies developed trying to inhibit it (de Streel and Lucas, 2021; Kim et al., 2021a). Initial attempts involved the use of neutralizing mAbs against the cytokine or its receptors (Kim et al., 2021a), such as anti-TFGβ clone

1D11, with many studies reporting an impact on NK cell function alone or when combined with other activating molecules (Alvarez et al., 2014, 2020b; Arteaga et al., 1993; Liu et al., 2017a). Fresolimumab (a humanized 1D11 mAb) neutralized TGFβ1, TGFβ2 and TGFβ 3 and has shown promising results in early clinical studies, but development has been discontinued in cancer (Morris et al., 2014) and pursued in profibrotic diseases. Interestingly, a new variant of fresolimumab, SAR439459, which displays a single mutation in the Fc region, has proved to not only inhibit the effects of anti-TGFβ on NK and T cells, but also to potentiate the efficacy of anti-PD-1 therapy on tumor mouse models (Greco et al., 2020; Kim et al., 2021a). Similarly, the highly selective mAb against TGFβ1, SRK181-mIgG1, has also shown to overcome the resistance to immune checkpoint blockade (ICB) therapy in experimental animals (Martin et al., 2020). The use of small molecule inhibitors against the TGF-β receptor kinase activity, and lately the development of TGF-β ligand traps are other means to neutralize TGF-β currently under investigation (Kim et al., 2021a). Among them, bintrafusp alfa (GSK-4045154, M7824, MSB0011359C), an innovative first-in-class bifunctional fusion protein composed of the mAb avelumab directed to PD-L1 fused to the extracellular domain of the TGF-β receptor as a soluble decoy. This agent is gaining attention due to promising early-phase results observed in several clinical trials (NCT02517398 (Cho et al., 2020), NCT02699515 (Kang et al., 2020), NCT02517398 (Paz-Ares et al., 2020)).

2.3.2 IL-10

IL-10 is another important cytokine that regulates NK cell activation (Ni et al., 2020; Rallis et al., 2021). However, the introduction of neutralizing IL-10 strategies in cancer therapy has been challenging due to the severe life-threating inflammatory toxicities (Rallis et al., 2021) as well as the role of IL-10 and IL-10R on the homeostasis and proliferation of immune cells (Ni et al., 2020). A summary of the key preclinical and clinical results obtained from the manipulation of IL-10 for cancer therapy has been published elsewhere (Ni et al., 2020). Pegilodecakin is a first-in-class, long-acting IL-10 receptor agonist that is being evaluated in several clinical trials alone or in combination with ICB therapy to enhance T cell function (NCT02923921 (Hecht et al., 2021), NCT02009449 (Naing et al., 2019), NCT02009449 (Tannir et al., 2021)). Although these studies did not evaluate the impact of pegilodecakin on NK cells, it is likely that NK cells would benefit from this agonist. IL-10-armed oncolytic viruses or engineered

IL-10 constructs to improve CAR activity *are other alternatives that are being explored to neutralize IL-10 and could easily be implemented towards NK cell-based immunotherapies*. However, some adverse events have been reported on patients receiving Pegilodecakin (Naing et al., 2016). Targeting IL-10 inhibition within the tumor site can be a way to prevent unwanted treatment-related side effects. This concept was conveyed by Qiao and colleges who created an anti-EGFR antibody (Cetuximab)-based IL-10 fusion protein (CmAb-(IL10) (Qiao et al., 2019). This molecule displayed enhanced anti-tumor responses on a EGFR$^+$ tumor mouse model, but more importantly, a reduction of toxicity was also *observed* (Qiao et al., 2019).

2.3.3 Inhibitory receptors

Neutralization of inhibitory receptors have also been considered for cancer immunotherapy to enhance NK cell function (Bi and Tian, 2019; Chauhan et al., 2020; Giuliani and Poggi, 2020; Myers and Miller, 2021; Sanchez-Correa et al., 2019; Zhang and Liu, 2020). The NKG2A-CD94 dimer is an inhibitory receptor whose expression levels become highly upregulated, as many inhibitory receptors do, once NK cells become activated (Alvarez et al., 2014, 2019), thus having an important role in controlling NK cell function. Tumor infiltrating NK cells have shown to express high levels of surface NKG2A and a correlation has been made between the frequency of NKG2A$^+$ NK cells secretion of IL-15 and TGFβ (Chauhan et al., 2020). Additionally, high levels of HLA-E, the human non-classical class I HLA NKG2A ligand, has been detected in many types of cancer, including hematological cancers (Giuliani and Poggi, 2020). Recent work has shown that systemic neutralization of the inhibitory receptor NKG2A resulted in strengthened anti-tumor responses by increasing NK and/or CD8 T cell effector functions that were further enhanced when combined with anti-PD-1 therapy (Andre et al., 2018; van Montfoort et al., 2018). There are eleven clinical studies that have evaluated or are evaluating the anti-tumor efficacy of monalizumab, a humanized anti-NKG2A immunoglobulin G (IgG) 4-blocking mAb, alone or in combination with other strategies that aim to increase NK cell-mediated ADCC and CD8 T cell effector functions. Thus far, results obtained from these clinical studies have indicated that monalizumab is well tolerated and more efficacious when combined with mAb targeting therapy (cetuximab or ibrutinib) and/or ICB therapy (NCT02643550 (Cohen et al., 2017; Fayette et al., 2018; Fayette et al., 2020; Tinker et al., 2019), NCT03088059 (Galot et al., 2021)). This corresponds with the context that the NKG2A inhibitory pathway might have a role in the resistance of PD-1 ICB therapy (Zhang et al., 2021a). HLA-E and its mouse orthologue, Qa-1b, present peptides in a TAP1-dependent manner to the NKG2A/CD94 heterodimer on both NK and CD8 T cells. It was shown that in the absence of TAP, tumors were resistant to anti-PD-1 treatment and Qa-1b could present an alternative peptidome to NKG2A$^+$ effector cells likely contributing towards a suppressive state. This resistance could be overcome when tumors were also deficient for Qa1b (Zhang et al., 2021a). Interestingly, CD94 can dimerize with the activatory receptor NKG2C, which also recognizes HLA-E. In this case NK cells express either NKG2A or NKG2C in mutually exclusive fashion.

TIGIT and CD96 are other inhibitory receptors of interest as their affinity for their ligands CD155 and CD112, which are found on many cancer types, are far more superior than the affinity of these stress molecules for the activating receptor DNAM-1 (Mittal et al., 2019; Yeo et al., 2021; Zhang et al., 2016a). Upregulation of both TIGIT and CD96 has been observed on NK cells exhausted through prolonged exposure of stimulatory signals (Alvarez et al., 2019, 2020c; Merino et al., 2019; Zhang et al., 2021b). In agreement with these studies, the expression of TIGIT was also found on dysfunctional NK cells form AML patients and correlated with poor prognosis (Liu et al., 2021a). Consequently, it was expected that the inhibition of TIGIT with mAbs caused strong NK and CD8 T cell-mediated anti-tumor responses alone or combined with other immunotherapeutic strategies (Hung et al., 2018; Judge et al., 2020b; Lozano et al., 2020; Mao et al., 2021). Many clinical trials are currently testing the efficacy of TIGIT neutralization through mAb therapy (tiragolumab, domvanalimab, etigilimab, ociperlimab, IBI939). Preliminary data from one of these trials suggest that tiragolumab, in combination with PD-L1 blockade, might be effective against solid cancers (NCT03563716 (Tiragolumab Impresses in Multiple Trials, 2020). A novel anti-TIGIT monoclonal antibody, AET2010 is also exhibiting excellent NK cell-mediated anti-tumor responses in preclinical studies (Han et al., 2021). Furthermore, SP8374, a fully human monoclonal immunoglobulin (Ig) G4 antibody designed to block the interaction of TIGIT with its ligands, has also promising results in preclinical settings (Shirasuna et al., 2021). The contribution of NK cells to TIGIT-based antigens has not been addressed yet.

With regard to CD96, the neutralization strategies based on mAb or other approaches are not as advanced as the ones for TIGIT are, but preclinical studies suggest that the blockade of CD96-CD155 interactions can better control metastasis in a NK cell-dependent manner (Roman Aguilera et al., 2018). A genetic and clinical characterization of CD96 on patients with glioblastoma showed a strong correlation of its expression with cancer progression (Liu et al., 2020a). Its expression was further associated with the expression of other inhibitory receptors such as TIGIT itself and thus its neutralization remains a promising therapeutic target (Liu et al., 2020a).

In addition to these inhibitory receptors, the neutralization of KIRs has also been explored for NK cell-based therapy. Lirilumab (IPH2102) is a second generation humanized antagonist antibody that targets the inhibitory KIR2DL1-3 and KIR2DS1-2 and is currently in clinical development, but preclinical studies have already shown promising results with enhanced NK cell-mediated cytotoxicity and ADCC against lymphoma, leukemia and multiple myeloma tumor cells (Benson Jr. et al., 2011; Kohrt et al., 2014; Romagne et al., 2009). In vitro data showed that lirilumab treatment synergized with daratumumab (anti-CD38) to increase NK cell-mediated ADCC on multiple myeloma tumor cell lines (NCT00552396 (Nijhof et al., 2015)). Although lirilumab was proven safe in a phase I clinical trial (Vey et al., 2018), a phase II study failed to demonstrate efficacy and development of this mAb has been terminated (NCT01248455 (Korde et al., 2014)). Looking at the preclinical studies, it is possible that the benefits of Lirilumab are better obtained when combined with other therapies that enhance NK cell activation.

2.3.4 Checkpoint inhibitors

TIM-3 and LAG-3 have also been considered for targeted therapy (Anderson et al., 2016). TIM-3 binds to galectin-9, phosphatidylserine, HMGB1, and CEACAM1. High levels of TIM-3 have been observed on NK cells from patients with non-small cell lung cancer, adenocarcinoma and gastrointestinal stromal tumors (Datar et al., 2019) and correlated with poor prognosis on AML and lung adenocarcinoma (Chajuwan et al., 2021; Wang et al., 2022; Xu et al., 2015). Its expression is frequently associated with NK cell dysfunction (Anderson et al., 2016; Yu et al., 2021). Inhibition of TIM-3 has resulted in enhanced NK cell cytotoxicity and the production of inflammatory cytokines (da Silva et al., 2014; Wang et al., 2015). Right now, there are several clinical trials that are exploring the impact of TIM-3 inhibition on patients for several cancers (Myers and Miller, 2021).

LAG-3, which binds to MHC class II and fibrinogen-like protein-1 (Wang et al., 2019), can regulate NK cell activation (Goldberg and

Drake, 2011). In fact, it has been shown that NK cells expressing LAG-3 display lower amounts of the inflammatory cytokines IFNγ and TNFα; and have impaired cytotoxic functions, but higher levels of IL-10 (Esen et al., 2021). There are multiple ongoing clinical studies that are investigating the impact of LAG-3 neutralization on effector T cells (Sanchez-Correa et al., 2019). However, unlike with T cells, there are no studies that have evaluated if LAG-3 neutralization improves NK cell effector functions. One would expect that LAG-3 inhibition might do so, maybe better when combined with the blockade of other inhibitory receptors and/or NK cell stimulating therapies (IL-15 or CD137 to name a few), which would be worthy of study (Myers and Miller, 2021; Sanchez-Correa et al., 2019).

One of the most successful histories for immunotherapy falls into the immunotherapeutic strategies that cause the inhibition of the PD-1/PD-L1 signaling pathway, which has revolutionized the way we see and treat cancer. The neutralization of PD-1 and/or PD-L1 have shown to restore T cell effector functions and overcome tolerance by delaying, preventing or reverting T cell exhaustion. Its remarkable anti-tumor efficacy in both preclinical and clinical studies, reviewed in-depth elsewhere (Chauhan et al., 2020; Zhang and Liu, 2020), led to the approval by the FDA and EMA of several agents like pembrolizumab, nivolumab and cemiplimab to target PD-1 and atezolizumab, avelumab and durvalumab to target PD-L1 (Twomey and Zhang, 2021). However, how this inhibitory signaling pathway affects NK cells function and exhaustion status is not clear, having studies suggesting a direct (Beldi-Ferchiou et al., 2016; Sun et al., 2020) or indirect impact (Alvarez et al., 2020c) depending on the models and types of cancer studied (Dunai and Murphy, 2018). According to Liu and collaborators, the number of PD-1$^+$ NK cells and its expression is increased in the TME of colon, liver, gastric, esophageal and ovarian cancer among many others (Liu et al., 2017b). In fact, PD-1 expression on NK cells from patients suffering different types of gastrointestinal cancers are correlated with poor prognosis (Myers and Miller, 2021). However, in other situations, low levels of PD-1 expression have been detected on NK cells (Alvarez et al., 2020c; Dunai and Murphy, 2018). Nevertheless, the consensus is that NK cell function is improved upon anti-PD-1/PD-L1 treatment (Alvarez et al., 2020c; Hsu et al., 2018; Sun et al., 2020). Interestingly, a role of NK cells in the anti-tumor efficacy of anti-PD-L1 treatment against PD-L1$^-$ tumors was recently suggested due to the ability of anti-PD-L1 to activate PD-L1-expressing NK cells (Dong et al., 2019), as PD-L1 surface expression is highly upregulated upon

NK cell activation (Alvarez et al., 2020c). An increase of NK cell-mediated ADDC was also observed on PD-L1$^+$ tumors (Julia et al., 2018). Although the main mechanisms for anti-PD-1/PD-L1 therapeutic efficacy has been attributed to effector T cells, nothing precludes NK cells from benefiting from a less immunosuppressive environment, directly or indirectly, contributing thus to the anti-PD-1/PD-L1 therapeutic effect through the cross-talk between NK cells, DCs and effector T cells (Bodder et al., 2020; Jacobs et al., 2021).

It is important to highlight that the current trends on NK cell-based therapies are advocating for combination therapies that involve immunotherapeutic strategies that modulate the effects of checkpoint inhibitors, PD-1/PDL-1 being the most common. However molecules that target other inhibitory receptors (TIGIT, NKG2A) or suppressor molecules (Cish, CD39, A2AR, CD73) are being considered for combination as well (Brauneck et al., 2021; Delconte et al., 2016; Neo et al., 2020). Not only that, lately many preclinical and clinical studies are surging where multiple immunosuppressor molecules are being simultaneously targeted (Brauneck et al., 2021; Mettu et al., 2021; Niu et al., 2021; Ravi et al., 2018). Such is the case for a first-in-human phase one study (NCT02964013) that is investigating the efficacy of anti-TIGIT combined with anti-PD-1 therapy in advance solid tumors (Niu et al., 2021). Moreover, a novel bispecific nanobody that consists of tetravalent anti-PD-L1 Fc-fusion nanobody and tetravalent anti-TIGIT nanobody is underdevelopment and has already demonstrated to increase the T cell anti-tumor efficacy (Ma et al., 2020). On the other hand, an in vitro study has also showed increased NK cell cytotoxicity against AML tumor cells when TIGIT and CD39 or A2AR were neutralized (Brauneck et al., 2021). Similarly, a study that evaluated the anti-tumor efficacy of bintrafusp alfa combined with NHS-IL12 (engineered by genetically fusing two human IL-12 heterodimers to the C-termini of the heavy chains of a human IgG1 antibody against DNA/Histone NHS76 that targets exposed dsDNA) demonstrated enhanced anti-tumor activity in preclinical tumor models by modulating both the activation and inhibition of NK cells (Xu et al., 2022).

3. NK-based cellular therapy
3.1 NK cells in hematopoietic stem cell transplantation

The potential benefits of NK cells in the setting of hematopoietic stem cell transplantation (HSCT) have long been recognized and utilized in the clinical setting. NK cells contribute to the graft-*versus*-leukemia (GVL) effect, which can be enhanced when part of the NK cell lacks inhibitory KIRs for HLA molecules expressed by the tumor, a situation termed KIR mismatch (Ames and Murphy, 2014). Paradoxically, NK cells are protective against graft-*versus*-host disease (GVHD), which occurs when donor alloreactive T cells attack host tissues. The mechanism by which NK cells suppress GVHD is complex; however, NK cell alloreactivity against either recipient antigen–presenting cells (APCs), alloreactive T cells, or both contribute to the effect (Simonetta et al., 2017). The indirect attack of alloreactive APCs has been demonstrated through suppression of mixed lymphocyte reaction (MLR) assays and by flow cytometry-based assays showing decreased viability of allogeneic DCs after culture with NK cells (Meinhardt et al., 2015). NK cells have also been shown to directly attack alloreactive T cell clones, via NK cell recognition of NKG2D ligands which are upregulated after T cells become activated (Olson et al., 2010).

In an important 2002 study, Ruggeri et al. compared the outcomes of HLA haploidentical HSCT recipients with KIR-ligand incompatibility in the graft-*versus*-host direction (Ruggeri et al., 2002). Those patients demonstrating "anti-recipient alloreactive NK clones" showed decreased rates of acute GVHD, rejection and among patients with AML, decreased probability of relapse at 5 years post-transplant. In a non-HSCT setting, Miller et al. tested haploidentical infusions of NK cells in patients with poor-prognosis AML, revealing an *in vivo* expansion of NK cells and complete hematologic remission in a subset of patients (Miller et al., 2005). In contrast, prior studies examining autologous transfers of NK cells showed no definite anti-tumor effect (Burns et al., 2003), further highlighting the importance of KIR mismatch with recipient/tumor HLA. In post-HSCT settings, NK cell infusions are generally well tolerated (Simonetta et al., 2017); however, one report in which *ex vivo* activated NK cells were administered at the time of engraftment following HLA-matched, T cell-depleted non-myeloablative HSCT showed development of acute GVHD in five out of nine patients (Shah et al., 2015). Though the mechanism by which these patients developed acute GVHD is unclear, the timing NK cells were infused may have exacerbated residual alloreactive T cells in the graft, possibly through the production of NK-derived cytokines, including IFN-γ.

3.2 Chimeric antigen receptor (CAR) and engineered NK cells

Adoptive cell therapies utilizing T cells engineered express a chimeric antigen receptor (CAR-T) are becoming established in clinical practice for

B-cell malignancies and show an improving safety profile (Lakshman and Kumar, 2021). At the time of this report, five CAR-T therapies have gained FDA approval for use in B-ALL, CLL, B-cell lymphomas and more recently multiple myeloma. Key obstacles impeding more widespread clinical use of CAR-T cells include high manufacturing costs, patient ineligibility, toxicity (namely cytokine release syndrome [CRS] and neurotoxicity (Safarzadeh Kozani et al., 2021)) and resistance/antigen loss (Lemoine et al., 2021; Plaks et al., 2021).

The high manufacturing cost and long development times for CAR-T based therapies are commanded by the requirement for an individualized/autologous production for each patient. Thus, a key advantage of an NK cell-based platform is their applicability in an allogeneic setting, as NK cells do not specifically require HLA-matching for their efficacy and are not known to cause directly GVHD. Similar to CAR-T cells, CAR-NK cells are *ex vivo* genetically modified NK cells that express a synthetic receptor construct with an extracellular antigen-binding domain that targets a specific tumor associated antigens (TAAs) via an antibody scFv, typically directed towards hematological malignancies TAAs, although efforts to implement CAR-NK cell-based therapy are being made towards solid tumors. In addition, it also encloses an extracellular domain linked to intracellular ITAM-containing signaling domain that typically involve CD28, 41BB (CD137), OX40 and DAP12/NKG2D among others (Marofi et al., 2021; Myers and Miller, 2021). Potential sources of so called "off-the-shelf" NK cell products include mature NK cells from cell lines, peripheral blood, or umbilical cord blood (UCB) or differentiating NK cells from immature precursors, such as from induced pluripotent stem cells (iPSCs). Importantly, recent data from MD Anderson investigators has demonstrated a remarkable clinical activity of UCB-derived allogeneic CD19 CAR-NK cells in ALL with seven out 11 patients achieving complete remission without treatment-related adverse events such as cytokine syndrome or GvHD (Liu et al., 2020b). Giving these exiting results, ensuing studies are under way.

3.2.1 Cell line-derived engineered NK cells
NK cell lines have been a popular source of NK cells for an "off-the-shelf" model of CAR-NK. A key advantage of a cell line-based therapy is the limitless replicative potential of the product and tolerance to freeze/thaw cycles, which permits manufacturing of large batches and allowing for a more standardized product and quality profile. Among the known NK cell lines, including YT, NKG, NKL, and others (Klingemann et al., 2016),

NK-92 remains the most studied cell line and the only one to advance to clinical trials. The NK-92 cell line was derived from a patient with an unspecified non-Hodgkin lymphoma and circulating large granular lymphocytes. Importantly, NK-92 maintain their cytotoxicity *in vitro* and *in vivo*, making them an appealing model for NK-based immunotherapies (Klingemann et al., 2016). However, this cell line-based strategy and NK-92 in particular present significant disadvantages limiting their use, especially when compared with donor-derived NK cells. First, NK-92 necessitate irradiation prior to transfusion to mitigate their malignant potential, thereby preventing a potentially desirable *in vivo* expansion. Next, NK-92 lack surface expression of CD16 (Gong et al., 1994), limiting potential synergy with tumor-targeting mAb therapies via ADCC. However, in its favor, NK-92 are highly amenable to genetic engineering (Gong et al., 2021) and have been investigated in preclinical CAR-NK models with engineered recognition of CD3 (Chen et al., 2016a), CD5 (Chen et al., 2017; Xu et al., 2019), CD7 (You et al., 2019), CD19 (Boissel et al., 2013; Oelsner et al., 2017; Romanski et al., 2016), CD20 (Boissel et al., 2013; Muller et al., 2008), CD38 (Hambach et al., 2020), CD138 (Luanpitpong et al., 2021), BCMA (Ng et al., 2021), CS1 (Chu et al., 2014), EBNA (Tassev et al., 2012), EGFR (Chen et al., 2016b; Genssler et al., 2016; Han et al., 2015), EGFRvIII (Genssler et al., 2016; Han et al., 2015), EpCAM (Zhang et al., 2018), GD2 (Esser et al., 2012), HER2 (Nowakowska et al., 2018; Zhang et al., 2016b) and mesothelin (Cao et al., 2020), among others.

Though the unmodified NK-92 cell line has been investigated in clinical trials for decades, trials of CAR-NK derived from NK-92 or other cell lines have been limited. A trial of CD33-specific CAR-NK showed no significant adverse events at does up to 5×10^9 cells/patient in patients with relapsed/refractory AML (Tang et al., 2018). Additional trials involving BCMA-specific CAR-NK 92 cells (multiple myeloma; NCT03940833) and HER2 CAR-NK (glioblastoma; NCT03383978) are actively recruiting. Finally, a trial investigating so-called chimeric costimulatory converting receptor (CCCR) NK cells, comprised of the extracellular portion of PD1, transmembrane and cytoplasmic domains of NKG2D and cytoplasmic domain of 41BB/CD137 represents a fascinating new paradigm for NK-based therapies (Lu et al., 2020a). Though not strictly a CAR-NK by definition, this approach aims to redirect the inhibitory signaling of PD-1 into a potent activation signal for the transfused product. This trial (NCT03656705) is enrolling by invitation at the time of publication. More recently, a group has developed a NK-92-derived novel NK cell line (oNK-1) with endogenous expression of CD16, higher expression of activating receptors and low expression of inhibitory receptors that was also conjugated to trastuzumab (Li et al., 2021a). The final product, named ACE1702, could eliminate HER2$^+$ cancer cells (Li et al., 2021a) and it is currently being evaluated in a phase I clinical trial for advanced and metastatic HER2-expressing solid tumors (NCT04319757).

3.2.2 Umbilical cord blood (UCB)-derived engineered NK cells

UCB offers a relatively accessible source of NK cells with a known safety profile, especially given the established use of UCB as a source of hematopoietic stem cells in HSCT, often combining multiple cord donors for a single recipient. Though limited by the number of NK cells that can be isolated from a single unit, UCB-derived NK cells have been shown to be capable of logarithmic expansion (Liu et al., 2018). UCB-derived NK cells are immature; though their proportion of CD56$^{bright-}$ and CD56^{dim} subsets is equivocal when compared with adult peripheral blood derived NK cells; UCB-NK cells express decreased KIRs, NKG2C, NKp46, DNAM-1, cytotoxic granule proteins perforin and granzyme B, as well as the death receptor ligand FasL (Luevano et al., 2012). Unstimulated UCB-derived NK cells show poor cytolytic activity, but can be induced by stimulation with IL-2. Similarly, UCB-derived NK cells show significantly lower levels of IFN-γ secretion upon stimulation when compared to those derived from adult PBMCs.

A notable trial involving the use of anti-CD19 engineered UCB-derived NK cells reported a response in eight of 11 patients with B cell lymphomas and chronic lymphocytic leukemia (CLL) (Liu et al., 2020b). The NK cell product in these patients was present nearly a year after the initial infusion, though interestingly, the relative abundance of the CAR-NK cells present beyond 14 days was not dependent on the initial dose of product. Some patients demonstrated a remarkable *in vivo* expansion of CAR-NK cells, though no recipient or donor factors were readily identified to predict NK cells persistence or expansion. An IL-15 construct in the engineered product was included to enhance the survival of the infused CAR-NK product, as unmodified and cytokine activated NK cells typically persist only several weeks post-transfusion (Miller et al., 2005). Thus, UCB-derived engineered NK cells have established themselves as a viable source for "off-the-shelf" models of CAR-NK cells. The vigorous expansion of NK cells from a single cord unit, on the order of 2000-fold or greater (Liu et al., 2018), permits the use of a single cord for over 100 doses of

CAR-NK cells. This allows the number of donor cords to undergo the manufacturing process to be minimized, thus reducing manufacturing costs and lead-time in developing the product.

3.2.3 Stem cell-derived engineered NK cells

The most significant step towards a truly homogeneous and renewable "off-the-shelf" engineered NK cell product is likely to be the use of human embryonic stem cells (hESC) or induced pluripotent stem cells (iPSC) as an NK cell source. The use of readily available cell lines, for iPSCs in particular, reduces the inherent issues of donor heterogeneity, efficiency of construct transduction, and variability in fold-expansion of the product, when compared with other NK cell sources. Therapies involving iPSC-derived cells have not shown tumorigenicity of the final product, thus abrogating the need for irradiation of the NK cells (Ueda et al., 2020). In contrast to cell line or UCB-derived sources, NK cells isolated and expanded from iPSCs also show normal expression of activating receptors, including CD16, when compared with PBMC-derived NK cells (Knorr et al., 2013) an potentially can be used to enhance ADCC of tumor-targeting mAb therapy. Though the process of manufacturing NK cells from an iPSC source is slower and more complex than from other sources, this process can theoretically be scaled and completed in large batches.

Indeed, a flurry of iPSC-derived NK cell therapies have entered clinical trials over the past several years. A series of engineered NK cell products from Fate Therapeutics (San Diego, CA, USA) have been applied to a broad range of hematologic and non-hematologic malignancies. These include FT538, an iPSC-derived NK cell therapy with three engineered modalities: a non-cleavable CD16, IL-15/IL-15 receptor alpha fusion and CD38 knockout. FT538 is being assessed in in patients with relapsed/refractory multiple myeloma or AML in combination with the monoclonal antibodies daratumumab (human IgG1 anti-CD38) or elotuzumab (humanized IgG1 anti-SLAMF7) (Bjordahl et al., 2019). FT538 is essentially a master of ADCC, especially in combination with daratumumab as the lack of CD38 expression on the product confers resistance to antibody-mediated fratricide.

A similar derived product, FT596 also expresses a non-cleavable CD16, a recombinant fusion of IL-15 and IL-15 receptor alpha as well as a CD19-direct CAR. The CD19 chimeric receptor is combined with a transmembrane domain of NKG2D and intracellular domains of CD3ζ, thereby tailored for NK cell effector functions (Therapeutics, 2021). A dose escalation study of FT596 has shown promising response rates in patients with relapsed refractory B cell lymphomas or CLL. This study had two cohorts, with patients receiving either a single dose of CAR-NK cells at various doses, or CAR-NK cells in combination with rituximab. When the cohorts were examined together, 10/14 patients achieved an objective response, with seven having a complete response. Interestingly, two of four patients who had received prior CD19-direct CAR-T therapy achieved a complete response (Bachanova et al., 2020).

3.2.4 CRISPR/Cas9-based engineered NK cells

The implementation of CRISPR/Cas9 technology has no escaped NK cell-based immunotherapy to support and facilitate the engineering of NK cells in order to increase their activation/function and/or prevent suppression (Afolabi et al., 2019; Grote et al., 2021; Huang et al., 2020; Morimoto et al., 2021; Zhu et al., 2020). Such approach has been recently carried on by Morimoto and colleagues who generated TIM3 knockout (KO) NK cells that showed superior NK cell cytotoxicity against glioblastoma tumor cells (Morimoto et al., 2021). The same approach can be used on cell line-derived NK cells to obtain top-quality "off-the-shelf" NK cells (Grote et al., 2021; Huang et al., 2020; Zhu et al., 2020). CAR NK-92 cells targeting CD276 (B7-H3) along with CRISPR/Cas9-induced disruption of the inhibitory receptor NKG2A were generated to increase NK cell cytolysis against melanoma cell lines (Grote et al., 2021). Although the disruption of NKG2A did not increase the *in vitro* cytotoxicity of CD276-CAR NK92, this study shows the promising potential of CRISPR/Cas9 technology to genetically modify NK cells at different levels. Similarly, CRISPR/Cas9 technology was used to construct UCB-derived anti-CD19 CAR-NK cells with a deletion of cytokine-inducible srchomology-2-containing protein (CIS), which is a negative regulator of IL-15 signaling, and an insert for the constitutive IL-15 expression (Daher et al., 2021). These modifications increased the CAR-NK persistency and anti-tumor responses on lymphoma xenografts without meaningful off-target toxicities. No clinical trials that explored the use of CRIPR/Cas9 on NK cells have started as of yet, but it will be a matter of time as 15 different CAR-T cells with modifications based on CRIPR/Cas9 have entered clinical trials, with just two trials currently completed for esophageal cancer (NCT03081715) and lung cancer (NCT02793856). In this open-label single-arm phase I clinical trial,

PD-1 KO CRISPR-edited T cells have proven to be safe and feasible for patients with refractory non-small-cell lung cancer (NCT02793856, (Lu et al., 2020b)).

An alternative approach for the generation of CARs has also taken advantage of the mRNA technology to genetically modified NK cells (Robbins et al., 2021). Xiao et al. adopted an NKG2D RNA CAR approach to significantly enhance the cytolytic activity of NK cells against several solid tumors (Xiao et al., 2019). This preclinical data supported the initiation of a clinical trial (NCT03415100) to evaluate the safety and feasibility of NKG2D CAR-NK cell therapy (Robbins et al., 2021; Xiao et al., 2019).

4. NK immune cell engagers

As a way to improve tumor targeting, similar to what has been done with T cell-based therapy, multiple strategies have been or are under development, which the direct modification of NK cells, like CAR-NKs; but others are aimed to take advantage of mAb-based therapies that rely on modified antibodies to induce ADCC, as is the case for NK immune cell engagers (NKCE). These molecules can be bi-specific (BiKE), tri-specific (TriKE) or even tetra-specific (TetraKE) killer engager proteins composed of multiple antibody-derived targeting domains; typically, single-chain variable fragments (scFvs). These NKCEs aim to increase specific tumor-targeting by the presence of scFvs that binds to a particular TAA (CD33, CD19, CD20, CD133, EpCAM, HER2, etc.). They also exploit NK cell-mediated ADCC with the addition of domains that trigger NK cell activation through CD16, or other activating receptors such as NKp46 or NKG2D (Phung et al., 2021), facilitating the formation of immunological synapses between tumors and NK cells (Felices et al., 2016).

CD16 BiKEs showed efficacy by increasing NK cell-mediated killing against tumor cells expressing CD19, CD33, CD133 or EpCAM (Hodgins et al., 2019; Schmohl et al., 2016b). However, TriKEs offer the advantage of a better therapeutic option when compared to BiKEs by also improving the maturation, proliferation, survival and homeostasis of NK cells with the addition of a functional significant IL-15 linker (Schmohl et al., 2016a). It was shown that TriKEs which incorporate anti-CD16 and anti-EpCAM scFvs linked with an IL-15 moiety resulted in better cytolytic activity against EpCAM$^+$ tumors, inflammatory cytokine production and proliferation of NK cells when compared to EpCAM BiKE (Schmohl et al., 2016a). Similar results have been obtained for a CD33-targeting

161533 TriKE (Yun et al., 2018), CD19-targeting 161519 TriKE (Cheng et al., 2020; Felices et al., 2019) or CD133-targeting 1615133 TriKE (Schmohl et al., 2017). However, in addition to these aspects of NK cell biology, 163315 TriKE, also named GTB-3550, was shown to improve the migratory abilities of NK cells and the NK cell serial killing capability; and decrease the time needed for the first tumor cell killing (Sarhan et al., 2018). These results prompted the entry of GTB-3550 (GT Biopharma) to an open-label phase I/II clinical trial for high-risk hematological malignances (NCT03214666). Although the results of this study are yet to be reported, according to Dr. Miller and Dr. Felices, the infusion of GTB-3550 has accomplished a robust NK cell expansion in all treated patients and the dose tested has been well-tolerated (Phung et al., 2021).

Interestingly, during the development of first generation 161533 TriKEs, it was noted that the use of wild type IL-15 as a linker between the two scFv arms resulted in a nonfunctional IL-15 and thus in this first generation a mutant IL-15 version with a substitution on N72D was chosen. A recent study suggested that the steric hindrance between the two scFv arms caused an inefficient folding that impaired the functionality of wild type IL-15 used as a linker (Felices et al., 2020). Therefore, in the second generation 161533 TriKE^{TM}, a modification of the anti-CD16 scFv with a humanized single-domain antibody against CD16 has enabled the presentation of functional wild type IL-15 and seems to be more effective than first generation TriKEs (Felices et al., 2020). Moreover, preclinical studies using a second generation CLEC12A (Arvindam et al., 2021) or HER2 (Vallera et al., 2021) TriKE^{TM} have recently demonstrate promising results against AML and ovarian cancer respectively.

Clearly, the versatility of NK cell immune engagers presents an attractive therapeutic avenue. Because of it, multiple pharmaceutical companies (GT BioPharma, Affimed, Dragonfly Therapeutics, Innate Pharma, Cytovia Therapeutics, Compass Therapeutics) have concentrated their efforts in developing their own NKCE versions (TriKE^{TM}, ROCK$^{®}$, TriNKET^{TM}, ANKET^{TM}, FLEX-NK$^{®}$) to target different tumors, many of which have already shown promising anti-tumor responses and are currently under study in clinical trials (Phung et al., 2021). The redirected optimized cell killing platform (ROCK$^{®}$) from Affimed was the first to test its products (AFM13, AFM24) on clinical trials and currently there are four active clinical trials for these two compounds (NCT04074746, NCT04259450, NCT05109442, NCT05099549,). AFM13 is a first-in-class tetravalent bispecific anti-CD30/CD16A antibody that consist in two diabodies with scFV domains against CD16A and CD30 (Wu et al., 2015). A phase I study (NCT0122157) showed that AFM13 was well-tolerated and safe in patients with relapse or refractory Hodgkin lymphoma (Rothe et al., 2015). From 26 patients, three achieved partial remission (11.5%), while 13 achieved stable disease (50%) with an overall disease control rate of 61.5%. A phase 1b dose-escalation study that involved AFM13 and pembrolizumab (NCT02665650) was also well-tolerated and displayed similar safety profiles than AFM13 alone, resulting in a 88% objective response rate at the highest treatment dose, with an 83% overall response rate for the overall population in patients with relapse or refractory Hodgkin lymphoma (Bartlett et al., 2020). Additionally, a recent published preclinical study demonstrated that the combination of AFM13 with cytokine-activated NK cells further increased the anti-tumor responses *in vitro* and *in vivo* against $CD30^+$ lymphomas (Kerbauy et al., 2021). Likewise, the preclinical evaluation of AFM24, an EGFR/CD16A tetravalent bispecific antibody, has also showed the tremendous potential of ROCK® platform against $EGFR^+$ tumors (Wingert et al., 2021) and it is currently being evaluated as monotherapy in a phase 1/2a open-label clinical trial for solid tumor patients (NCT04259450). FMA24 is also been evaluated in combination with atezolizumab (NCT05109442) or autologous non-genetically modified NK cells SNK01 (NKGen Biotech) (NCT05099549). AFM28 targeting CD123 and AFM32 are currently under development for acute myeloid leukemia and solid tumors respectively.

Unlike the other approaches, TriNKET™ platform seems to rely on NKG2D to exert the NK cell activation, been the DF1001 the first TriKE engager from Dragonfly Therapeutics to be evaluated in the clinic for the treatment of advance solid $HER2^+$ tumors in a phase I/2 clinical trial alone or in combination with the ICB nivolumab or paclitaxel (NCT04143711).

On the other hand, Innate Pharma has developed an NKp46/CD16-based NK cell engager (IPH6101/SAR443579) using an antibody-based NK cell engager therapeutics (ANKET™) platform (Gauthier et al., 2019). IPH6101 have shown NK cell-mediated superior anti-tumor activity in preclinical models and is currently being tested in a first-in-human clinical trial in patients with relapsed or refractory hematological malignances (NCT05086315). NKp46 is also used in the FLEX-NK® (Cytovia therapeutics). Currently, Cytovia Therapeutics is developing several NK cell engagers that involve CYT303, CYT100 and CYT150 targeting GPC3 for hepatocarcinoma and solid tumors, CYT338 and CYT539 targeting CD38 for multiple myeloma and CYT501 targeting EGFR for glioblastoma and solid tumors, but none of their product have entered clinical evaluation just yet. Alternatively, Compass Therapeutics is working on a novel multi-specific NK cell engager with anti-CD16A, anti-NKp30 and anti-BCMA targeting domains, CTX-8573, that has shown strong NK and γδ T cells-mediated anti-tumor responses both *in vivo* and *in vitro* thanks to superior ADCC lytic capacities displayed by these effector cells (Watkins-Yoon et al., 2019). This approach could represent a promising alternative for the treatment of multiple myeloma patients because unlike daratumumab (anti-CD38), CTX-8573 does not induce NK cell fratricide.

Most interestingly, NKCEs have also shown to have benefits on the persistence and cytotoxic potential of CAR-NK cells. Zhang and collaborators recently demonstrated that a bispecific antibody, which harbors an N-terminal scFv antibody domain for binding to NKG2D, linked via a human IgG4 Fc region to a second C-terminal scFv antibody domain for recognition of ErbB2, was able not only to increase the lysis of Erb2$^+$ tumor cells by peripheral blood-derived NK cells; but also could synergize with CAR-NK92 cells engineered to express an NKG2D-CD3ζ chimeric antigen receptor (Zhang et al., 2021c). This approach should prevent the negative impact of soluble NKG2DLs and endogenous NKG2D down-regulation that are known to impair NK cell function (Alvarez et al., 2019; Zhang et al., 2021c). Additionally, this work endorses future clinical studies that combine NKCEs with CAR-NK cell therapy.

5. Oncolytic virotherapy

Oncolytic virotherapy (OVT) is an emerging cancer treatment, which takes advantage of the direct cytotoxic lytic capacities from effector cells and the following immune-mediated response generated upon viral infection. Some of the viruses that have been used for OVT involve adenovirus, herpes virus, reovirus and poxivirus, among others, and currently there are multiple preclinical and clinical research ongoing (Zhao et al., 2021). Oncolytic virus (OV) are characterized by their ability to selectively infect and eliminate malignant cells due to the expression of viral entry receptors on these cells (Leung and McNeish, 2021). Such is the case for the expression of CD46 for adenovirus (Zhao et al., 2021) and measles virus (Anderson et al., 2004) on multiple myeloma (Ong et al., 2006) and many other cancers

(Anderson et al., 2004); or CD155 and the integrin α2β1 molecules for polyvirus and echovirus respectively (Anderson et al., 2004; Bergelson et al., 1992). Alternatively, OV can also be engineered to improve tumor tropism, enhance tumor selectivity and/or increase anti-tumor efficacy (Kangas et al., 2021; Zhao et al., 2021). Some of these modifications include alterations to selectively infect tumor cells through the recognition of surface receptors or TAAs highly present on malignant cells like EGFR (Aghi et al., 2008; Coffey et al., 1998; Leung and McNeish, 2021; Uusi-Kerttula et al., 2018), carcinoembryonic antigen (CEA) (Gatti-Mays et al., 2020; Sun et al., 2019), mucin 1 (MUC1) or brachyury for colorectal cancer (Gatti-Mays et al., 2020), among others.

How OVT works has been reviewed elsewhere in detail (Boagni et al., 2021; Kangas et al., 2021; Kazemi et al., 2021; Li et al., 2021b; Mealiea and McCart, 2021). Briefly, once the malignant cells are infected by the OV, subsequent viral replication causes direct oncolysis and the release of tumor-specific antigens that unleash an immune response primarily mediated by DCs and cytotoxic effector cells. Concurrently, OVs can promote immunogenic cell death (ICD) that results in the release of viral pathogen-associated and damage-associated molecular patterns (PAMPs and DAMPs respectively) that are sensed by the pattern recognition receptors (PRRs) present on many immune cells like toll-like receptors (TLRs), RIG-1 or cGAs. Triggering these receptors induces the activation of multiple inflammatory signaling pathways like protein kinase RNA-activated (PKR), TANK binding kinase 1 (TBK-1) and/or stimulator of interferon genes (STING), all of which increase type I IFN production and NF-κB signaling, and contribute to the recruitment and maturation of DCs, as well as the expansion of antigen-specific effector cells (Ferrucci et al., 2021; Kangas et al., 2021; Kazemi et al., 2021; Mealiea and McCart, 2021). Because of these changes, TME becomes highly inflamed and it has been suggested that OVT can promote the conversion of what is called "cold" unresponsive non-inflamed tumors into "hot" responsive inflamed tumors due to the increase of proinflammatory cytokines and chemokines and the recruitment of effector cells, NK cells among them (Hodgins et al., 2019; Marotel et al., 2020).

Currently, three OVs have been approved for the treatment of solid tumors. An unmodified Enteric Cytopathogenic Human Orphan Type 7 (ECHO-7) picornavirus (Rigvir) and Talimogene laherparepvec (T-VEC, Imlygic) are approved for the treatment of unresectable metastatic melanoma, and a recombinant adenovirus (Oncorine) is approved for the treatment of head and neck tumors (Kazemi et al., 2021).

Several studies have pointed out the importance of NK cells in the efficacy of OVT (Altomonte et al., 2008; Alvarez-Breckenridge et al., 2012; Annels et al., 2021; Hodgins et al., 2019; Liu et al., 2021b; Mealiea and McCart, 2021; Medina-Echeverz et al., 2019; Miller and Fraser, 2003; Miyamoto et al., 2012; Ogbomo et al., 2010; Wongthida et al., 2010; Zhang et al., 2021d). On one hand, some studies have suggested that the presence of NK cells can negatively affect OVT due to direct clearance of OV-infected cells, preventing the viral replication and spread and thus precluding OVT from working (Altomonte et al., 2008; Alvarez-Breckenridge et al., 2012). *An vitro* study using a 3D spheroid cultures showed that oncolytic parainfluenza virus 5-infected cancer cells were efficiently killed by NK cells in a NKp30, NKp46 and NKG2D-dependent manner and correlated with interferon induction (Varudkar et al., 2021). Similarly, several studies have shown that the inhibition of NK and NKT cell function through the blockage of activating receptors (DNAM-1, NKp46 or NKp30) enhances the oncolytic potency of OVT (Altomonte et al., 2008; Alvarez-Breckenridge et al., 2012; Bar-On et al., 2017). On the other hand, in many studies an increase of NK cell recruitment and activation has been frequently observed after OVT and has been correlated with the OVT efficacy (Alkayyal et al., 2017; Dempe et al., 2012; Ramelyte et al., 2021; Schwaiger et al., 2017; Zhang et al., 2014, 2020). Trying to solve this dichotomy, two mathematical models have exposed the relevance for the timing of NK cell presence in the TME to impair or support OVT anti-tumor efficacy (Kim et al., 2018; Senekal et al., 2021). According to these studies, early NK cell presence seems to be detrimental for the OVT efficacy, likely due to the eradication of OV-infected cells (Kim et al., 2018; Senekal et al., 2021). However, the recruitment of NK cells to the TME later on, once the OVT has started, does not only enhance anti-tumor responses (Kim et al., 2018; Senekal et al., 2021), but it seems to be critical for OVT outcomes (Dempe et al., 2012; Kim et al., 2018; Ramelyte et al., 2021; Senekal et al., 2021). Additionally, it has been shown that OVT can enhance NK-mediated cytotoxic potential by upregulating natural cytotoxicity and activating receptors (Bhat et al., 2011; Klose et al., 2019).

Taking together, these studies advocate for combination therapies that involve OVT and NK cells. In fact, numerous attempts have been made to increase OVT-mediated NK cell dependent anti-tumor responses. Some of these strategies include engineer OVs to express NK cell-stimulating cytokines IL-12 (Alkayyal et al., 2017; Backhaus et al., 2019; Choi et al., 2012; Quetglas et al., 2015), IL-15 (Backhaus et al., 2019; Niu et al., 2015; Stephenson et al., 2012; Tosic et al., 2014), IL-18 (Zheng et al., 2009), CCL5 (Lapteva et al., 2009; Li et al., 2020) or GM-CSF (Andtbacka et al., 2015; Ramelyte et al., 2021), or combination of cytokines; such is the case for TRAIL/IL-12 (El-Shemi et al., 2016), IL12/IL-18(Choi et al., 2011), IL-12/IL-18/B7.1 (Fukuhara et al., 2005; Ino et al., 2006), GM-CSF/IL-2 (Tian et al., 2021) or GM-CSF/IL-12 (Kim et al., 2021b). Modified OVs armed with CD137 (Thomas et al., 2019), CTLA-4 (Thomas et al., 2019) and/or PD-1 (Lin et al., 2020; Thomas et al., 2019; Tian et al., 2021; Xie et al., 2022) have also been studied. A recent and different approach has consisted in increasing the ADCC and antibody-dependent cellular phagocytosis (ADCP) by NK cells and macrophages respectively, using an oncolytic herpes virus modified to express a full-length anti(α)-human CD47 IgG1 or IgG4 antibody and preventing therefore the CD47-SIRPα checkpoint inhibition (Xu et al., 2021).

Recently, other strategies aim to exploit OVT by combining it with NK cell-based adoptive transfer therapies as it is expected that both therapies complement and help each other by favoring OVs infection on one hand and increasing the recruitment and activation of NK cells on the other hand (Chen et al., 2016b; Gao et al., 2020; Han et al., 2015; Klose et al., 2019; Li et al., 2020; Ma et al., 2021). In this regard, a recent study has shown that OV-expressing human IL15/IL15Rα sushi domain fusion protein (OV-IL15C) plus "off-the-shelf" EGFR-CAR NK cells augmented the presence of NK and CD8 T cells in the TME and increased the survival and cytotoxic potential of these cells (Ma et al., 2021). Moreover, CAR-NK cells profited from OVT with a superior survival rate (Ma et al., 2021). Consequently, this strategy resulted in a potent anti-tumor response in an immunocompetent multiple glioblastoma mouse model when compared to OV-IL15C or CAR-NK monotherapies (Ma et al., 2021). Similarly, the intratumoral injections of CCL-5-expressing vaccinia virus on a HCT-116 tumor xenograft mouse model increased the homing to the TME of intravenously injected NK cells engineered to express the CCL5 ligand CCR5. As expected, this combination resulted in a better control of tumor growth and thus, the tumor free survival of the treated mice was heightened (Li et al., 2020). Recognizing the potential of this combinatorial approach, it would not be unrealistic that clinical studies exploiting these two therapeutic interventions start soon and nothing precludes evaluating OVT therapy with other successful NK cell-based therapies such as BiKEs or TriKES as it is being currently done for BiTEs (Speck et al., 2018).

6. Conclusions

Pre-clinical and clinical research on the therapeutic exploitation of NK cells is ongoing and holds promise. Decades since their discovery, their potent effector activities, which complement those of T cells, keep them in the forefront of cancer research. Understanding the mechanisms involved in the regulation of this complex immune cell has been critical for the progress of new immunotherapeutic strategies to elicit potent and sustained anti-tumor responses against cancer. Thanks to this knowledge, it is clear now that triggering just NK cell activation is barely enough to keep NK cell function and thus many approaches are currently being under development and based on modulating multiple aspects on NK cell regulation; both activators and suppressors (Fig. 1). In fact, if we take a look into current clinical trials (as of February 2022) that aim to modulate NK cell function at some level, combination therapies that regulate NK cells at multiple levels are clearly endorsed (Fig. 2).

Following this line of thought, combination of NKCE approaches, that trigger NK cell activation and facilitate tumor targeting, with ICB therapy are currently being explored in several clinical trials (NCT02665650, NCT05109442, NCT04143711). Similarly, CAR-NK cell-based therapy is also evaluating its efficacy in combination with ICB (NCT03228667, NCT04050709, NCT04390399, and NCT04847466). Interestingly, the infusion of PDL1-targeted high-affinity NK cells (PD-L1 t-haNK), an "off-the-shelf" NK cell line derived from the NK-92 cells engineered to express high-affinity CD16, endoplasmic reticulum-retained interleukin IL-2, and a PD-L1-specific chimeric antigen receptor (CAR), is being tested in combination with the IL-15 superagonist N-803 along with (NCT03228667, NCT04847466) or without (NCT04390399) ICB therapy. Multiple clinical trials at different stages are also studding the efficacy of pembrolizumab and ALKS 4230 (Nemvaleukin Alfa), an engineered fusion protein comprised of a circularly-permuted IL-2 with the extracellular domain of IL-2Rα, for a variety of cancers (NCT04592653, NCT03861793, NCT02799095, NCT04144517, NCT05092360, NCT04830124). In addition to ICB therapy, combining NK cell-based therapy with strategies

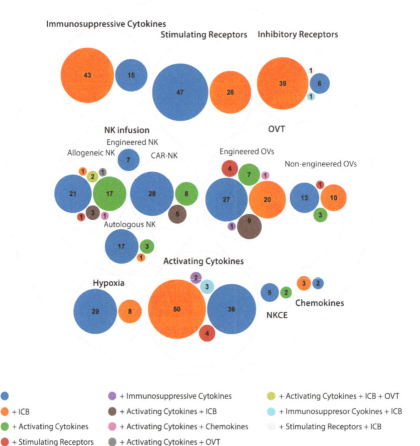

Fig. 2 Current landscape of clinical trials that modulate NK cell activation and function. The figure shows the clinical trials that as of February 2022 are not yet recruiting, recruiting or active for strategies that target some immunosuppressive cytokines (TGFβ and IL-10), activating cytokines (IL-2, IL-12, IL-15 and IL-18), inhibitory receptors (KIRs, NKG2A, CD96 and TIGIT), stimulating receptors (CD137, OX40, TNFR2, SLAM7), chemokines (CCR1/2, CCR4, CCR5), hypoxia molecules (ADAR2, CD73, CD39) and oncolytic virotherapy (OVT) that are known to influence NK cells; as well as trials that involve the infusion of engineered or non-engineered NK cells derived from different sources. It does highlight the current trends towards combined therapies (represented with different colors) that aim to shape NK cells at multiple levels, having special relevance the therapies that are combined with ICB therapy (PD-(L)1, TIM3, LAG-3, CTLA-4). The dark blue circle in each subgroup represents clinical trials given as monotherapy or combined with other therapies not specified in the figure such as targeting-mAb therapy like trastuzumab or cetuximab depending of the type of cancer, to name a few. For simplicity, the use of radiotherapy or chemotherapeutic drugs has not been taking under consideration. The number under each circle represents the total number of clinical trials for that particular section. *Figure created with RAWGraphs 2.0 beta (https://doi.org/10.1145/3125571.3125585).*

to prevent suppression by blocking inhibitory receptors and/or suppressor cytokines could also be considered. As an example for this approach, a phase I clinical trials has been initiated for patients with recurrent glioblastoma to evaluate the safety of UCB-derived NK cells containing deleted TGF-betaR2 and NR3C1 (CB-NK-TGF-betaR2$^-$/NR3C1$^-$) that make them unresponsive to TGFβ and glucocorticoids (NCT04991870). Another phase 1b study will evaluate the safety and activity of ex vivo preactivated and expanded CB-NK-TGF-betaR2$^-$/NR3C1$^-$ cells in combination with cetuximab for colorectal cancer patients (NCT05040568). All of these trials represent good examples of where NK cell-based therapy is moving towards and their results are eagerly awaited so the true potential of these promising innate immune cells can be proven.

Acknowledgments

M.F.S. is funded by the Navarra Government project LINTERNA (Ref.: 0011-1411-2020-000075). F.S. is funded by the *Ligue genevoise contre le cancer* (LGC 2011). MA is funded by the Spanish Association Against Cancer's Investigator grant (INVES19041ALVA). The authors would like to thank Dr. Pedro Berraondo and CD, EA and MA would like to specially acknowledge Dr. William J. Murphy for his NK cell and Ph.D. mentorship.

Conflict of interest

MA reports receiving research funding from Highlight Therapeutics and PharmaMar. The rest of the authors have no conflict of interest to declare.

References

Afolabi, L.O., Adeshakin, A.O., Sani, M.M., Bi, J., Wan, X., 2019. Genetic reprogramming for NK cell cancer immunotherapy with CRISPR/Cas9. Immunology 158 (2), 63–69. PubMed PMID: 31315144. Pubmed Central PMCID: 6742769.

Aghi, M., Visted, T., Depinho, R.A., Chiocca, E.A., 2008. Oncolytic herpes virus with defective ICP6 specifically replicates in quiescent cells with homozygous genetic mutations in p16. Oncogene 27 (30), 4249–4254. PubMed PMID: 18345032. Pubmed Central PMCID: 7100519.

Alderson, K.L., Sondel, P.M., 2011. Clinical cancer therapy by NK cells via antibody-dependent cell-mediated cytotoxicity. J. Biomed. Biotechnol. 2011, 379123. PubMed PMID: 21660134. Pubmed Central PMCID: 3110303.

Alkayyal, A.A., Tai, L.H., Kennedy, M.A., de Souza, C.T., Zhang, J., Lefebvre, C., et al., 2017. NK-cell recruitment is necessary for eradication of peritoneal carcinomatosis with an IL12-expressing Maraba virus cellular vaccine. Cancer Immunol. Res. 5 (3), 211–221. PubMed PMID: 28159747.

Altomonte, J., Wu, L., Chen, L., Meseck, M., Ebert, O., Garcia-Sastre, A., et al., 2008. Exponential enhancement of oncolytic vesicular stomatitis virus potency by vector-mediated suppression of inflammatory responses in vivo. Mol. Ther. 16 (1), 146–153. PubMed PMID: 18071337. Pubmed Central PMCID: 2930752.

Alvarez, M., Bouchlaka, M.N., Sckisel, G.D., Sungur, C.M., Chen, M., Murphy, W.J., 2014. Increased antitumor effects using IL-2 with anti-TGF-beta reveals competition between mouse NK and CD8 T cells. J. Immunol. 193 (4), 1709–1716. PubMed PMID: 25000978. Pubmed Central PMCID: 4241855.

Alvarez, M., Simonetta, F., Baker, J., Pierini, A., Wenokur, A.S., Morrison, A.R., et al., 2019. Regulation of murine NK cell exhaustion through the activation of the DNA damage repair pathway. JCI Insight 18, 5. PubMed PMID: 31211693. Pubmed Central PMCID: 6675585.

Alvarez, M., Pierini, A., Simonetta, F., Baker, J., Maas-Bauer, K., Hirai, T., et al., 2020a. Infusion of host-derived unlicensed NK cells improves donor engraftment in non-myeloablative allogeneic hematopoietic cell transplantation. Front. Immunol. 11, 614250. PubMed PMID: 33488624. Pubmed Central PMCID: 7817981.

Alvarez, M., Dunai, C., Khuat, L.T., Aguilar, E.G., Barao, I., Murphy, W.J., 2020b. IL-2 and anti-TGF-beta promote NK cell reconstitution and anti-tumor effects after syngeneic hematopoietic stem cell transplantation. Cancer 12 (11). PubMed PMID: 33138229. Pubmed Central PMCID: 7692743.

Alvarez, M., Simonetta, F., Baker, J., Morrison, A.R., Wenokur, A.S., Pierini, A., et al., 2020c. Indirect impact of PD-1/PD-L1 blockade on a murine model of NK cell exhaustion. Front. Immunol. 11, 7. PubMed PMID: 32117218. Pubmed Central PMCID: 7026672.

Alvarez, M., Ochoa, M.C., Minute, L., Melero, I., Berraondo, P., 2020d. Rapid isolation and enrichment of mouse NK cells for experimental purposes. Methods Enzymol. 631, 257–275. PubMed PMID: 31948551.

Alvarez-Breckenridge, C.A., Yu, J., Price, R., Wojton, J., Pradarelli, J., Mao, H., et al., 2012. NK cells impede glioblastoma virotherapy through NKp30 and NKp46 natural cytotoxicity receptors. Nat. Med. 18 (12), 1827–1834. PubMed PMID: 23178246. Pubmed Central PMCID: 3668784.

Ames, E., Murphy, W.J., 2014. Advantages and clinical applications of natural killer cells in cancer immunotherapy. Cancer Immunol. Immunother. 63 (1), 21–28. PubMed PMID: 23989217. Pubmed Central PMCID: PMC3880590. Epub 2013/08/31.

Anderson, B.D., Nakamura, T., Russell, S.J., Peng, K.W., 2004. High CD46 receptor density determines preferential killing of tumor cells by oncolytic measles virus. Cancer Res. 64 (14), 4919–4926. PubMed PMID: 15256464.

Anderson, A.C., Joller, N., Kuchroo, V.K., 2016. lag-3, tim-3, and tigit: co-inhibitory receptors with specialized functions in immune regulation. Immunity 44 (5), 989–1004. PubMed PMID: 27192565. Pubmed Central PMCID: 4942846.

Andre, P., Denis, C., Soulas, C., Bourbon-Caillet, C., Lopez, J., Arnoux, T., et al., 2018. Anti-NKG2A mAb is a checkpoint inhibitor that promotes anti-tumor immunity by unleashing both T and NK cells. Cell 175 (7), 1731–1743. e13. PubMed PMID: 30503213. Pubmed Central PMCID: 6292840.

Andtbacka, R.H., Kaufman, H.L., Collichio, F., Amatruda, T., Senzer, N., Chesney, J., et al., 2015. Talimogene Laherparepvec improves durable response rate in patients with advanced melanoma. J. Clin. Oncol. Off. J. Am. Soc. Clin. Oncol. 33 (25), 2780–2788. PubMed PMID: 26014293.

Anfossi, N., Andre, P., Guia, S., Falk, C.S., Roetynck, S., Stewart, C.A., et al., 2006. Human NK cell education by inhibitory receptors for MHC class I. Immunity 25 (2), 331–342. PubMed PMID: 16901727.

Annels, N.E., Simpson, G.R., Denyer, M., Arif, M., Coffey, M., Melcher, A., et al., 2021. Oncolytic reovirus-mediated recruitment of early innate immune responses reverses immunotherapy resistance in prostate tumors. Mol. Ther. Oncolytics 20, 434–446. PubMed PMID: 33665363. Pubmed Central PMCID: 7900644.

Arteaga, C.L., Hurd, S.D., Winnier, A.R., Johnson, M.D., Fendly, B.M., Forbes, J.T., 1993. Anti-transforming growth factor (TGF)-beta antibodies inhibit breast cancer cell tumorigenicity and increase mouse spleen natural killer cell activity. Implications for a possible role of tumor cell/host TGF-beta interactions in human breast cancer progression. J. Clin. Invest. 92 (6), 2569–2576. PubMed PMID: 7504687. Pubmed Central PMCID: 288452.

Arvindam, U.S., van Hauten, P.M.M., Schirm, D., Schaap, N., Hobo, W., Blazar, B.R., et al., 2021. A trispecific killer engager molecule against CLEC12A effectively induces NK-cell mediated killing of AML cells. Leukemia 35 (6), 1586–1596. PubMed PMID: 33097838. Pubmed Central PMCID: 8189652.

Aste-Amezaga, M., D'Andrea, A., Kubin, M., Trinchieri, G., 1994. Cooperation of natural killer cell stimulatory factor/interleukin-12 with other stimuli in the induction of cytokines and cytotoxic cell-associated molecules in human T and NK cells. Cell. Immunol. 156 (2), 480–492. PubMed PMID: 7912999.

Bachanova, V., Cayci, Z., Lewis, D., Maakaron, J.E., Janakiram, M., Bartz, A., et al., 2020. Initial clinical activity of FT596, a first-in-class, multi-antigen targeted, off-the-shelf, iPSC-derived CD19 CAR NK cell therapy in relapsed/refractory b-cell lymphoma. Blood 136. PubMed PMID: WOS:000607547203153. English.

Backhaus, P.S., Veinalde, R., Hartmann, L., Dunder, J.E., Jeworowski, L.M., Albert, J., et al., 2019. Immunological effects and viral gene expression determine the efficacy of oncolytic measles vaccines encoding IL-12 or IL-15 agonists. Viruses 11 (10). PubMed PMID: 31623390. Pubmed Central PMCID: 6832518.

Barao, I., Hanash, A.M., Hallett, W., Welniak, L.A., Sun, K., Redelman, D., et al., 2006. Suppression of natural killer cell-mediated bone marrow cell rejection by CD4+CD25+ regulatory T cells. Proc. Natl. Acad. Sci. U. S. A. 103 (14), 5460–5465. PubMed PMID: 16567639. Pubmed Central PMCID: 1459377.

Barao, I., Alvarez, M., Ames, E., Orr, M.T., Stefanski, H.E., Blazar, B.R., et al., 2011. Mouse Ly49G2+ NK cells dominate early responses during both immune reconstitution and activation independently of MHC. Blood 117 (26), 7032–7041. PubMed PMID: 21498673. Pubmed Central PMCID: 3143551.

Bar-On, Y., Charpak-Amikam, Y., Glasner, A., Isaacson, B., Duev-Cohen, A., Tsukerman, P., et al., 2017. NKp46 recognizes the sigma1 protein of reovirus: implications for reovirus-based cancer therapy. J. Virol. 91 (19). PubMed PMID: 28724773. Pubmed Central PMCID: 5599737.

Bartlett, N.L., Herrera, A.F., Domingo-Domenech, E., Mehta, A., Forero-Torres, A., Garcia-Sanz, R., et al., 2020. A phase 1b study of AFM13 in combination with pembrolizumab in patients with relapsed or refractory Hodgkin lymphoma. Blood 136 (21), 2401–2409. PubMed PMID: 32730586. Pubmed Central PMCID: 7685206.

Beck, J.D., Reidenbach, D., Salomon, N., Sahin, U., Tureci, O., Vormehr, M., et al., 2021. mRNA therapeutics in cancer immunotherapy. Mol. Cancer 20 (1), 69. PubMed PMID: 33858437. Pubmed Central PMCID: 8047518.

Bekaii-Saab, T.S., Roda, J.M., Guenterberg, K.D., Ramaswamy, B., Young, D.C., Ferketich, A.K., et al., 2009. A phase I trial of paclitaxel and trastuzumab in combination with interleukin-12 in patients with HER2/neu-expressing malignancies. Mol. Cancer Ther. 8 (11), 2983–2991. PubMed PMID: 19887543. Pubmed Central PMCID: 2996611.

Beldi-Ferchiou, A., Lambert, M., Dogniaux, S., Vely, F., Vivier, E., Olive, D., et al., 2016. PD-1 mediates functional exhaustion of activated NK cells in patients with Kaposi sarcoma. Oncotarget 7 (45), 72961–72977. PubMed PMID: 27662664. Pubmed Central PMCID: 5341956.

Benson Jr., D.M., Bakan, C.E., Zhang, S., Collins, S.M., Liang, J., Srivastava, S., et al., 2011. IPH2101, a novel anti-inhibitory KIR antibody, and lenalidomide combine to enhance the natural killer cell versus multiple myeloma effect. Blood 118 (24), 6387–6391. PubMed PMID: 22031859. Pubmed Central PMCID: 3490103.

Bergelson, J.M., Shepley, M.P., Chan, B.M., Hemler, M.E., Finberg, R.W., 1992. Identification of the integrin VLA-2 as a receptor for echovirus 1. Science 255 (5052), 1718–1720. PubMed PMID: 1553561.

Berrien-Elliott, M.M., Becker-Hapak, M., Cashen, A.F., Jacobs, M.T., Wong, P., Foster, M., et al., 2021. Systemic IL-15 promotes allogeneic cell rejection in patients treated with natural killer cell adoptive therapy. Blood. PubMed PMID: 34797911.

Bessard, A., Sole, V., Bouchaud, G., Quemener, A., Jacques, Y., 2009. High antitumor activity of RLI, an interleukin-15 (IL-15)-IL-15 receptor alpha fusion protein, in metastatic melanoma and colorectal cancer. Mol. Cancer Ther. 8 (9), 2736–2745. PubMed PMID: 19723883.

Bhat, R., Dempe, S., Dinsart, C., Rommelaere, J., 2011. Enhancement of NK cell antitumor responses using an oncolytic parvovirus. Int. J. Cancer 128 (4), 908–919. PubMed PMID: 20473905.

Bi, J., Tian, Z., 2019. NK cell dysfunction and checkpoint immunotherapy. Front. Immunol. 10, 1999. PubMed PMID: 31552017. Pubmed Central PMCID: 6736636.

Bjordahl R, Gaidarova S, Woan K, Cichocki F, Bonello G, Robinson M, et al. FT538: preclinical development of an off-the-shelf adoptive NK cell immunotherapy with targeted disruption of CD38 to prevent anti-CD38 antibody-mediated fratricide and enhance ADCC in multiple myeloma when combined with daratumumab. Blood 2019;134 (Supplement_1):133-.

Boagni, D.A., Ravirala, D., Zhang, S.X., 2021. Current strategies in engaging oncolytic viruses with antitumor immunity. Mol. Ther. Oncolytics 22, 98–113. PubMed PMID: 34514092. Pubmed Central PMCID: 8411207. oncolytic herpes simplex virus constructed in his lab for clinical application.

Bodder, J., Zahan, T., van Slooten, R., Schreibelt, G., de Vries, I.J.M., Florez-Grau, G., 2020. Harnessing the cDC1-NK cross-talk in the tumor microenvironment to battle cancer. Front. Immunol. 11, 631713. PubMed PMID: 33679726. Pubmed Central PMCID: 7933030.

Boieri, M., Ulvmoen, A., Sudworth, A., Lendrem, C., Collin, M., Dickinson, A.M., et al., 2017. IL-12, IL-15, and IL-18 pre-activated NK cells target resistant T cell acute lymphoblastic leukemia and delay leukemia development in vivo. Oncoimmunology 6 (3), e1274478. PubMed PMID: 28405496. Pubmed Central PMCID: 5384344.

Boissel, L., Betancur-Boissel, M., Lu, W., Krause, D.S., Van Etten, R.A., Wels, W.S., et al., 2013. Retargeting NK-92 cells by means of CD19- and CD20-specific chimeric antigen receptors compares favorably with antibody-dependent cellular cytotoxicity. Oncoimmunology 2 (10), e26527. PubMed PMID: 24404423. Pubmed Central PMCID: PMC3881109. Epub 2014/01/10.

Brauneck, F., Seubert, E., Wellbrock, J., Schulze Zur Wiesch, J., Duan, Y., Magnus, T., et al., 2021. Combined blockade of TIGIT and CD39 or A2AR enhances NK-92 cell-mediated cytotoxicity in AML. Int. J. Mol. Sci. 22 (23). PubMed PMID: 34884723. Pubmed Central PMCID: 8657570.

Burns, L.J., Weisdorf, D.J., DeFor, T.E., Vesole, D.H., Repka, T.L., Blazar, B.R., et al., 2003. IL-2-based immunotherapy after autologous transplantation for lymphoma and breast cancer induces immune activation and cytokine release: a phase I/II trial. Bone Marrow Transplant. 32 (2), 177–186. PubMed PMID: 12838283. Epub 2003/07/03.

Cabo, M., Santana-Hernandez, S., Costa-Garcia, M., Rea, A., Lozano-Rodriguez, R., Ataya, M., et al., 2021. CD137 costimulation counteracts TGFbeta inhibition of NK-cell antitumor function. Cancer Immunol. Res. 9 (12), 1476–1490. PubMed PMID: 34580116.

Cao, B., Liu, M., Wang, L., Liang, B., Feng, Y., Chen, X., et al., 2020. Use of chimeric antigen receptor NK-92 cells to target mesothelin in ovarian cancer. Biochem. Biophys. Res. Commun. 524 (1), 96–102. PubMed PMID: 31980173. Epub 2020/01/26.

Carlsten, M., Levy, E., Karambelkar, A., Li, L., Reger, R., Berg, M., et al., 2016. Efficient mRNA-based genetic engineering of human NK cells with high-affinity CD16 and CCR7 augments Rituximab-induced ADCC against lymphoma and targets NK cell migration toward the lymph node-associated chemokine CCL19. Front. Immunol. 7, 105. PubMed PMID: 27047492. Pubmed Central PMCID: 4801851.

Chajuwan, T., Kansuwan, P., Kobbuaklee, S., Chanswangphuwana, C., 2021. Characteristics and clinical correlation of TIM-3 and PD-1/PD-L1 expressions in leukemic cells and tumor microenvironment in newly diagnosed acute myeloid leukemia. Leuk. Lymphoma 29, 1–7. PubMed PMID: 34585994.

Chambers, A.M., Matosevic, S., 2019. Immunometabolic dysfunction of natural killer cells mediated by the hypoxia-CD73 axis in solid tumors. Front. Mol. Biosci. 6, 60. PubMed PMID: 31396523. Pubmed Central PMCID: 6668567.

Chauhan, S.K.S., Koehl, U., Kloess, S., 2020. Harnessing nk cell checkpoint-modulating immunotherapies. Cancer 12 (7). PubMed PMID: 32640575. Pubmed Central PMCID: 7408278.

Chen, K.H., Wada, M., Firor, A.E., Pinz, K.G., Jares, A., Liu, H., et al., 2016a. Novel anti-CD3 chimeric antigen receptor targeting of aggressive T cell malignancies. Oncotarget 7 (35), 56219–56232. PubMed PMID: 27494836. Pubmed Central PMCID: PMC5302909. Epub 2016/08/06.

Chen, X., Han, J., Chu, J., Zhang, L., Zhang, J., Chen, C., et al., 2016b. A combinational therapy of EGFR-CAR NK cells and oncolytic herpes simplex virus 1 for breast cancer brain metastases. Oncotarget 7 (19), 27764–27777. PubMed PMID: 27050072. Pubmed Central PMCID: 5053686.

Chen, K.H., Wada, M., Pinz, K.G., Liu, H., Lin, K.W., Jares, A., et al., 2017. Preclinical targeting of aggressive T-cell malignancies using anti-CD5 chimeric antigen receptor. Leukemia 31 (10), 2151–2160. PubMed PMID: 28074066. Pubmed Central PMCID: PMC5629371. Epub 2017/01/12.

Cheng, Y., Zheng, X., Wang, X., Chen, Y., Wei, H., Sun, R., et al., 2020. Trispecific killer engager 161519 enhances natural killer cell function and provides anti-tumor activity against CD19-positive cancers. Cancer Biol. Med. 17 (4), 1026–1038. PubMed PMID: 33299651. Pubmed Central PMCID: 7721099.

Chertova, E., Bergamaschi, C., Chertov, O., Sowder, R., Bear, J., Roser, J.D., et al., 2013. Characterization and favorable in vivo properties of heterodimeric soluble IL-15. IL-15Ralpha cytokine compared to IL-15 monomer. J. Biol. Chem. 288 (25), 18093–18103. PubMed PMID: 23649624. Pubmed Central PMCID: 3689953.

Cho, B.C., Daste, A., Ravaud, A., Salas, S., Isambert, N., McClay, E., et al., 2020. Bintrafusp alfa, a bifunctional fusion protein targeting TGF-beta and PD-L1, in advanced squamous cell carcinoma of the head and neck: results from a phase I cohort. J. Immunother. Cancer 8 (2). PubMed PMID: 32641320. Pubmed Central PMCID: 7342865.

Choi, I.K., Lee, J.S., Zhang, S.N., Park, J., Sonn, C.H., Lee, K.M., et al., 2011. Oncolytic adenovirus co-expressing IL-12 and IL-18 improves tumor-specific immunity via differentiation of T cells expressing IL-12Rbeta2 or IL-18Ralpha. Gene Ther. 18 (9), 898–909. PubMed PMID: 21451575. Pubmed Central PMCID: 3169807.

Choi, K.J., Zhang, S.N., Choi, I.K., Kim, J.S., Yun, C.O., 2012. Strengthening of antitumor immune memory and prevention of thymic atrophy mediated by adenovirus expressing IL-12 and GM-CSF. Gene Ther. 19 (7), 711–723. PubMed PMID: 21993173.

Chu, J., Deng, Y., Benson, D.M., He, S., Hughes, T., Zhang, J., et al., 2014. CS1-specific chimeric antigen receptor (CAR)-engineered natural killer cells enhance in vitro and in vivo antitumor activity against human multiple myeloma. Leukemia 28 (4), 917–927. PubMed PMID: 24067492. Pubmed Central PMCID: PMC3967004. Epub 2013/09/27.

Chu, D.T., Bac, N.D., Nguyen, K.H., Tien, N.L.B., Thanh, V.V., Nga, V.T., et al., 2019. An update on anti-CD137 antibodies in immunotherapies for cancer. Int. J. Mol. Sci. 20 (8). PubMed PMID: 31013788. Pubmed Central PMCID: 6515339.

Cirella, A., Berraondo, P., Di Trani, C.A., Melero, I., 2020. Interleukin-12 message in a bottle. Clin. Cancer Res. 26 (23), 6080–6082. PubMed PMID: 33004432.

Coffey, M.C., Strong, J.E., Forsyth, P.A., Lee, P.W., 1998. Reovirus therapy of tumors with activated Ras pathway. Science 282 (5392), 1332–1334. PubMed PMID: 9812900.

Cohen, R.B., Salas, S., Even, C., Kotecki, N., Jimeno, A., Soulie, A.M., et al., 2017. Safety of the first-in-class anti-NKG2A monoclonal antibody monalizumab in combination with cetuximab: a phase Ib/II study in recurrent or metastatic squamous cell carcinoma of the head and neck (R/M SCCHN). Cancer Res. PubMed PMID: WOS:000442513305276. English.

Conlon, K.C., Miljkovic, M.D., Waldmann, T.A., 2019. Cytokines in the treatment of cancer. J. Interferon Cytokine Res. 39 (1), 6–21. PubMed PMID: 29889594. Pubmed Central PMCID: 6350412.

Cooley, S., He, F., Bachanova, V., Vercellotti, G.M., DeFor, T.E., Curtsinger, J.M., et al., 2019. First-in-human trial of rhIL-15 and haploidentical natural killer cell therapy for advanced acute myeloid leukemia. Blood Adv. 3 (13), 1970–1980. PubMed PMID: 31266741. Pubmed Central PMCID: 6616260.

da Silva, I.P., Gallois, A., Jimenez-Baranda, S., Khan, S., Anderson, A.C., Kuchroo, V.K., et al., 2014. Reversal of NK-cell exhaustion in advanced melanoma by Tim-3 blockade. Cancer Immunol. Res. 2 (5), 410–422. PubMed PMID: 24795354. Pubmed Central PMCID: 4046278.

Daher, M., Basar, R., Gokdemir, E., Baran, N., Uprety, N., Nunez Cortes, A.K., et al., 2021. Targeting a cytokine checkpoint enhances the fitness of armored cord blood CAR-NK cells. Blood 137 (5), 624–636. PubMed PMID: 32902645. Pubmed Central PMCID: 7869185.

Datar, I., Sanmamed, M.F., Wang, J., Henick, B.S., Choi, J., Badri, T., et al., 2019. Expression analysis and significance of PD-1, LAG-3, and TIM-3 in human non-small cell lung cancer using spatially resolved and multiparametric single-cell analysis. Clin. Cancer Res. 25 (15), 4663–4673. PubMed PMID: 31053602. Pubmed Central PMCID: 7444693.

de Streel, G., Lucas, S., 2021. Targeting immunosuppression by TGF-beta1 for cancer immunotherapy. Biochem. Pharmacol. 192, 114697. PubMed PMID: 34302795. Pubmed Central PMCID: 8484859.

Delconte, R.B., Kolesnik, T.B., Dagley, L.F., Rautela, J., Shi, W., Putz, E.M., et al., 2016. CIS is a potent checkpoint in NK cell-mediated tumor immunity. Nat. Immunol. 17 (7), 816–824. PubMed PMID: 27213690.

Dempe, S., Lavie, M., Struyf, S., Bhat, R., Verbeke, H., Paschek, S., et al., 2012. Antitumoral activity of parvovirus-mediated IL-2 and MCP-3/CCL7 delivery into human pancreatic cancer: implication of leucocyte recruitment. Cancer Immunol. Immunother. CII 61 (11), 2113–2123. PubMed PMID: 22576056.

Di Trani, C.A., Fernandez-Sendin, M., Cirella, A., Segues, A., Olivera, I., Bolanos, E., et al., 2022. Advances in mRNA-based drug discovery in cancer immunotherapy. Expert Opin. Drug Discov. 17 (1), 41–53. PubMed PMID: 34496689.

Dogra, P., Rancan, C., Ma, W., Toth, M., Senda, T., Carpenter, D.J., et al., 2020. Tissue determinants of human NK cell development, function, and residence. Cell 180 (4), 749–763. e13. PubMed PMID: 32059780. Pubmed Central PMCID: 7194029.

Dong, W., Wu, X., Ma, S., Wang, Y., Nalin, A.P., Zhu, Z., et al., 2019. The mechanism of anti-PD-L1 antibody efficacy against PD-L1-negative tumors identifies NK cells expressing PD-L1 as a cytolytic effector. Cancer Discov. 9 (10), 1422–1437. PubMed PMID: 31340937. Pubmed Central PMCID: 7253691.

Dubois, S.P., Miljkovic, M.D., Fleisher, T.A., Pittaluga, S., Hsu-Albert, J., Bryant, B.R., et al., 2021. Short-course IL-15 given as a continuous infusion led to a massive expansion of effective NK cells: implications for combination therapy with antitumor antibodies. J. Immunother. Cancer 9 (4). PubMed PMID: 33883258. Pubmed Central PMCID: 8061813.

Dunai, C., Murphy, W.J., 2018. NK cells for PD-1/PD-L1 blockade immunotherapy: pinning down the NK cell. J. Clin. Invest. 128 (10), 4251–4253. PubMed PMID: 30198909. Pubmed Central PMCID: 6159962.

Elpek, K.G., Rubinstein, M.P., Bellemare-Pelletier, A., Goldrath, A.W., Turley, S.J., 2010. Mature natural killer cells with phenotypic and functional alterations accumulate upon sustained stimulation with IL-15/IL-15Ralpha complexes. Proc. Natl. Acad. Sci. U. S. A. 107 (50), 21647–21652. PubMed PMID: 21098276. Pubmed Central PMCID: 3003106.

El-Shemi, A.G., Ashshi, A.M., Na, Y., Li, Y., Basalamah, M., Al-Allaf, F.A., et al., 2016. Combined therapy with oncolytic adenoviruses encoding TRAIL and IL-12 genes markedly suppressed human hepatocellular carcinoma both in vitro and in an orthotopic transplanted mouse model. J. Exp. Clin. Cancer Res. CR 35, 74. PubMed PMID: 27154307. Pubmed Central PMCID: 4859966.

Esen, F., Deniz, G., Aktas, E.C., 2021. PD-1, CTLA-4, LAG-3, and TIGIT: the roles of immune checkpoint receptors on the regulation of human NK cell phenotype and functions. Immunol. Lett. 240, 15–23. PubMed PMID: 34599946.

Esser, R., Muller, T., Stefes, D., Kloess, S., Seidel, D., Gillies, S.D., et al., 2012. NK cells engineered to express a GD2-specific antigen receptor display built-in ADCC-like activity against tumour cells of neuroectodermal origin. J. Cell. Mol. Med. 16 (3), 569–581. PubMed PMID: 21595822. Pubmed Central PMCID: PMC3822932. Epub 2011/05/21.

Fayette, J., Lefebvre, G., Posner, M.R., Bauman, J., Salas, S., Even, C., et al., 2018. Results of a phase II study evaluating monalizumab in combination with cetuximab in previously treated recurrent or metastatic squamous cell carcinoma of the head and neck (R/M SCCHN). Ann. Oncol. 29, 374. PubMed PMID: WOS:000459277302227. English.

Fayette, J., Bauman, J., Salas, S., Colevas, D., Even, C., Cupissol, D., et al., 2020. Monalizumab in combination with cetuximab post platinum and anti-PD-(L)1 in patients with recurrent/metastatic squamous cell carcinoma of the head and neck (R/M SCCHN): updated results from a phase II trial. Ann. Oncol. 31. S1450-S. PubMed PMID: WOS:000600992500081. English.

Felices, M., Lenvik, T.R., Davis, Z.B., Miller, J.S., Vallera, D.A., 2016. Generation of BiKEs and TriKEs to improve NK cell-mediated targeting of tumor cells. Methods Mol. Biol. 1441, 333–346. PubMed PMID: 27177679. Pubmed Central PMCID: 5823010.

Felices, M., Lenvik, A.J., McElmurry, R., Chu, S., Hinderlie, P., Bendzick, L., et al., 2018. Continuous treatment with IL-15 exhausts human NK cells via a metabolic defect. JCI Insight 3 (3). PubMed PMID: 29415897. Pubmed Central PMCID: 5821201.

Felices, M., Kodal, B., Hinderlie, P., Kaminski, M.F., Cooley, S., Weisdorf, D.J., et al., 2019. Novel CD19-targeted TriKE restores NK cell function and proliferative capacity in CLL. Blood Adv. 3 (6), 897–907. PubMed PMID: 30890546. Pubmed Central PMCID: 6436008.

Felices, M., Lenvik, T.R., Kodal, B., Lenvik, A.J., Hinderlie, P., Bendzick, L.E., et al., 2020. Potent cytolytic activity and specific IL15 delivery in a second-generation trispecific killer engager. Cancer Immunol. Res. 8 (9), 1139–1149. PubMed PMID: 32661096. Pubmed Central PMCID: 7484162.

Ferrucci, P.F., Pala, L., Conforti, F., Cocorocchio, E., 2021. Talimogene Laherparepvec (T-VEC): an intralesional cancer immunotherapy for advanced melanoma. Cancer 13 (6). PubMed PMID: 33803762. Pubmed Central PMCID: 8003308.

Foster, J.B., Barrett, D.M., Kariko, K., 2019. The emerging role of in vitro-transcribed mRNA in adoptive T cell immunotherapy. Mol. Ther. 27 (4), 747–756. PubMed PMID: 30819612. Pubmed Central PMCID: 6453504.

Fujii, R., Jochems, C., Tritsch, S.R., Wong, H.C., Schlom, J., Hodge, J.W., 2018. An IL-15 superagonist/IL-15Ralpha fusion complex protects and rescues NK cell-cytotoxic function from TGF-beta1-mediated immunosuppression. Cancer Immunol. Immunother. CII 67 (4), 675–689. PubMed PMID: 29392336. Pubmed Central PMCID: 6326360.

Fukuhara, H., Ino, Y., Kuroda, T., Martuza, R.L., Todo, T., 2005. Triple gene-deleted oncolytic herpes simplex virus vector double-armed with interleukin 18 and soluble B7-1 constructed by bacterial artificial chromosome-mediated system. Cancer Res. 65 (23), 10663–10668. PubMed PMID: 16322208.

Furuya, H., Chan, O.T.M., Pagano, I., Zhu, C., Kim, N., Peres, R., et al., 2019. Effectiveness of two different dose administration regimens of an IL-15 superagonist complex (ALT-803) in an orthotopic bladder cancer mouse model. J. Transl. Med. 17 (1), 29. PubMed PMID: 30654801. Pubmed Central PMCID: 6337786.

Galot, R., Le Tourneau, C., Saada-Bouzid, E., Daste, A., Even, C., Debruyne, P., et al., 2021. A phase II study of monalizumab in patients with recurrent/metastatic squamous cell carcinoma of the head and neck: the I1 cohort of the EORTC-HNCG-1559 UPSTREAM trial. Eur. J. Cancer 158, 17–26. PubMed PMID: 34638090.

Gao, J., Zhang, W., Mese, K., Bunz, O., Lu, F., Ehrhardt, A., 2020. Transient chimeric Ad5/37 fiber enhances NK-92 carrier cell-mediated delivery of oncolytic adenovirus type 5 to tumor cells. Mol. Ther. Methods Clin. Dev. 18, 376–389. PubMed PMID: 32695840. Pubmed Central PMCID: 7358217.

Gatti-Mays, M.E., Redman, J.M., Donahue, R.N., Palena, C., Madan, R.A., Karzai, F., et al., 2020. A Phase I trial using a multitargeted recombinant adenovirus 5 (CEA/MUC1/Brachyury)-based immunotherapy vaccine regimen in patients with advanced cancer. Oncologist 25 (6), 479–e899. PubMed PMID: 31594913. Pubmed Central PMCID: 7288633.

Gauthier, L., Morel, A., Anceriz, N., Rossi, B., Blanchard-Alvarez, A., Grondin, G., et al., 2019. Multifunctional natural killer cell engagers targeting NKp46 trigger protective tumor immunity. Cell 177 (7), 1701–1713. e16. PubMed PMID: 31155232.

Gauthier, M., Laroye, C., Bensoussan, D., Boura, C., Decot, V., 2021. Natural Killer cells and monoclonal antibodies: two partners for successful antibody dependent cytotoxicity against tumor cells. Crit. Rev. Oncol. Hematol. 160, 103261. PubMed PMID: 33607229.

Genssler, S., Burger, M.C., Zhang, C., Oelsner, S., Mildenberger, I., Wagner, M., et al., 2016. Dual targeting of glioblastoma with chimeric antigen receptor-engineered natural killer cells overcomes heterogeneity of target antigen expression and enhances antitumor activity and survival. Oncoimmunology 5 (4), e1119354. PubMed PMID: 27141401. Pubmed Central PMCID: PMC4839317. Epub 2016/05/04.

Gillies, S.D., Lan, Y., Hettmann, T., Brunkhorst, B., Sun, Y., Mueller, S.O., et al., 2011. A low-toxicity IL-2-based immunocytokine retains antitumor activity despite its high degree of IL-2 receptor selectivity. Clin. Cancer Res. 17 (11), 3673–3685. PubMed PMID: 21531812.

Giuliani, M., Poggi, A., 2020. Checkpoint inhibitors and engineered cells: new weapons for natural killer cell arsenal against hematological malignancies. Cell 9 (7). PubMed PMID: 32610578. Pubmed Central PMCID: 7407972.

Goldberg, M.V., Drake, C.G., 2011. LAG-3 in cancer immunotherapy. Curr. Top. Microbiol. Immunol. 344, 269–278. PubMed PMID: 21086108. Pubmed Central PMCID: 4696019.

Gomez Garcia, L.M., Escudero, A., Mestre, C., Fuster Soler, J.L., Martinez, A.P., Vagace Valero, J.M., et al., 2021. Phase 2 clinical trial of infusing haploidentical K562-mb15-41BBL-activated and expanded natural killer cells as consolidation therapy for pediatric acute myeloblastic leukemia. Clin. Lymphoma Myeloma Leuk. 21 (5), 328–37 e1. PubMed PMID: 33610500.

Gong, J.H., Maki, G., Klingemann, H.G., 1994. Characterization of a human cell line (NK-92) with phenotypical and functional characteristics of activated natural killer cells. Leukemia 8 (4), 652–658. PubMed PMID: 8152260. Epub 1994/04/01.

Gong, Y., Klein Wolterink, R.G.J., Wang, J., Bos, G.M.J., Germeraad, W.T.V., 2021. Chimeric antigen receptor natural killer (CAR-NK) cell design and engineering for cancer therapy. J. Hematol. Oncol. 14 (1), 73. PubMed PMID: 33933160. Pubmed Central PMCID: PMC8088725. Epub 2021/05/03.

Greco, R., Qu, H., Qu, H., Theilhaber, J., Shapiro, G., Gregory, R., et al., 2020. Pan-TGFbeta inhibition by SAR439459 relieves immunosuppression and improves antitumor efficacy of PD-1 blockade. Oncoimmunology 9 (1), 1811605. PubMed PMID: 33224628. Pubmed Central PMCID: 7657645.

Greene, S., Robbins, Y., Mydlarz, W.K., Huynh, A.P., Schmitt, N.C., Friedman, J., et al., 2020. Inhibition of MDSC trafficking with SX-682, a CXCR1/2 inhibitor, enhances NK-cell immunotherapy in head and neck cancer models. Clin. Cancer Res. 26 (6), 1420–1431. PubMed PMID: 31848188. Pubmed Central PMCID: 7073293.

Grote, S., Urena-Bailen, G., Chan, K.C., Baden, C., Mezger, M., Handgretinger, R., et al., 2021. In vitro evaluation of CD276-CAR NK-92 functionality, migration and invasion potential in the presence of immune inhibitory factors of the tumor microenvironment. Cell 10 (5). PubMed PMID: 33925968. Pubmed Central PMCID: 8145105.

Guo, J., Liang, Y., Xue, D., Shen, J., Cai, Y., Zhu, J., et al., 2021a. Tumor-conditional IL-15 pro-cytokine reactivates anti-tumor immunity with limited toxicity. Cell Res. 31 (11), 1190–1198. PubMed PMID: 34376814. Pubmed Central PMCID: 8563767.

Guo, M., Sun, C., Qian, Y., Zhu, L., Ta, N., Wang, G., et al., 2021b. Proliferation of highly cytotoxic human natural killer cells by OX40L armed NK-92 with secretory neoleukin-2/15 for cancer immunotherapy. Front. Oncol. 11, 632540. PubMed PMID: 33937033. Pubmed Central PMCID: 8083131.

Hallett, W.H.D., Ames, E., Alvarez, M., Barao, I., Taylor, P.A., Blazar, B.R., et al., 2008. Combination therapy using IL-2 and anti-CD25 results in augmented natural killer cell-mediated antitumor responses. Biol. Blood Marrow Transplant. 14 (10), 1088–1099. PubMed PMID: 18804038. Pubmed Central PMCID: 2735407.

Hambach, J., Riecken, K., Cichutek, S., Schutze, K., Albrecht, B., Petry, K., et al., 2020. Targeting CD38-expressing multiple myeloma and Burkitt lymphoma cells in vitro with nanobody-based chimeric antigen receptors (Nb-CARs). Cell 9 (2). PubMed PMID: 32013131. Pubmed Central PMCID: PMC7072387. Epub 2020/02/06.

Han, J., Chu, J., Keung Chan, W., Zhang, J., Wang, Y., Cohen, J.B., et al., 2015. CAR-engineered NK cells targeting wild-type EGFR and EGFRvIII enhance killing of glioblastoma and patient-derived glioblastoma stem cells. Sci. Rep. 5, 11483. PubMed PMID: 26155832. Pubmed Central PMCID: 4496728.

Han, D., Xu, Y., Zhao, X., Mao, Y., Kang, Q., Wen, W., et al., 2021. A novel human anti-TIGIT monoclonal antibody with excellent function in eliciting NK cell-mediated antitumor immunity. Biochem. Biophys. Res. Commun. 534, 134–140. PubMed PMID: 33341068.

Hecht, J.R., Lonardi, S., Bendell, J., Sim, H.W., Macarulla, T., Lopez, C.D., et al., 2021. Randomized Phase III Study of FOLFOX alone or with pegilodecakin as second-line therapy in patients with metastatic pancreatic cancer that progressed after gemcitabine (SEQUOIA). J. Clin. Oncol. 39 (10), 1108–1118. PubMed PMID: 33555926. Pubmed Central PMCID: 8078437.

Herberman, R.B., Nunn, M.E., Lavrin, D.H., 1975. Natural cytotoxic reactivity of mouse lymphoid cells against syngeneic acid allogeneic tumors. I. Distribution of reactivity and specificity. Int. J. Cancer 16 (2), 216–229. PubMed PMID: 50294.

Hewitt, S.L., Bailey, D., Zielinski, J., Apte, A., Musenge, F., Karp, R., et al., 2020. Intratumoral IL12 mRNA therapy promotes TH1 transformation of the tumor microenvironment. Clin. Cancer Res. 26 (23), 6284–6298. PubMed PMID: 32817076.

Hodgins, J.J., Khan, S.T., Park, M.M., Auer, R.C., Ardolino, M., 2019. Killers 2.0: NK cell therapies at the forefront of cancer control. J. Clin. Invest. 129 (9), 3499–3510. PubMed PMID: 31478911. Pubmed Central PMCID: 6715409.

Hsu, J., Hodgins, J.J., Marathe, M., Nicolai, C.J., Bourgeois-Daigneault, M.C., Trevino, T.N., et al., 2018. Contribution of NK cells to immunotherapy mediated by PD-1/PD-L1 blockade. J. Clin. Invest. 128 (10), 4654–4668. PubMed PMID: 30198904. Pubmed Central PMCID: 6159991.

Huang, R.S., Shih, H.A., Lai, M.C., Chang, Y.J., Lin, S., 2020. Enhanced NK-92 cytotoxicity by CRISPR genome engineering using Cas9 ribonucleoproteins. Front. Immunol. 11, 1008. PubMed PMID: 32528479. Pubmed Central PMCID: 7256201.

Hung, A.L., Maxwell, R., Theodros, D., Belcaid, Z., Mathios, D., Luksik, A.S., et al., 2018. TIGIT and PD-1 dual checkpoint blockade enhances antitumor immunity and survival in GBM. Oncoimmunology. 7 (8), e1466769. PubMed PMID: 30221069. Pubmed Central PMCID: 6136875.

Ino, Y., Saeki, Y., Fukuhara, H., Todo, T., 2006. Triple combination of oncolytic herpes simplex virus-1 vectors armed with interleukin-12, interleukin-18, or soluble B7-1 results in enhanced antitumor efficacy. Clin. Cancer Res. 12 (2), 643–652. PubMed PMID: 16428511.

Jacobs, B., Gebel, V., Heger, L., Greze, V., Schild, H., Dudziak, D., et al., 2021. Characterization and manipulation of the crosstalk between dendritic and natural killer cells within the tumor microenvironment. Front. Immunol. 12, 670540. PubMed PMID: 34054844. Pubmed Central PMCID: 8160470.

Joncker, N.T., Fernandez, N.C., Treiner, E., Vivier, E., Raulet, D.H., 2009. NK cell responsiveness is tuned commensurate with the number of inhibitory receptors for self-MHC class I: the rheostat model. J. Immunol. 182 (8), 4572–4580. PubMed PMID: 19342631. Pubmed Central PMCID: 2938179.

Judge, S.J., Murphy, W.J., Canter, R.J., 2020a. Characterizing the dysfunctional NK cell: assessing the clinical relevance of exhaustion, anergy, and senescence. Front. Cell. Infect. Microbiol. 10, 49. PubMed PMID: 32117816. Pubmed Central PMCID: 7031155.

Judge, S.J., Darrow, M.A., Thorpe, S.W., Gingrich, A.A., O'Donnell, E.F., Bellini, A.R., et al., 2020b. Analysis of tumor-infiltrating NK and T cells highlights IL-15 stimulation and TIGIT blockade as a combination immunotherapy strategy for soft tissue sarcomas. J. Immunotherapy Cancer 8 (2). PubMed PMID: 33158916. Pubmed Central PMCID: 7651745.

Julia, E.P., Amante, A., Pampena, M.B., Mordoh, J., Levy, E.M., 2018. Avelumab, an IgG1 anti-PD-L1 immune checkpoint inhibitor triggers NK cell-mediated cytotoxicity and cytokine production against triple negative breast cancer cells. Front. Immunol. 9, 2140. PubMed PMID: 30294328. Pubmed Central PMCID: 6159755.

Kang, Y.K., Bang, Y.J., Kondo, S., Chung, H.C., Muro, K., Dussault, I., et al., 2020. Safety and tolerability of bintrafusp Alfa, a bifunctional fusion protein targeting TGFbeta and PD-L1, in Asian patients with pretreated recurrent or refractory gastric cancer. Clin. Cancer Res. 26 (13), 3202–3210. PubMed PMID: 32299818.

Kangas, C., Krawczyk, E., He, B., 2021. Oncolytic HSV: underpinnings of tumor susceptibility. Viruses 13 (7). PubMed PMID: 34372614. Pubmed Central PMCID: 8310378.

Kazemi, M.H., Kuhestani Dehaghi, B., Roshandel, E., Parkhideh, S., Mehdizadeh, M., Salimi, M., et al., 2021. Oncolytic virotherapy in hematopoietic stem cell transplantation. Hum. Immunol. 82 (9), 640–648. PubMed PMID: 34119352.

Kerbauy, L.N., Marin, N.D., Kaplan, M., Banerjee, P.P., Berrien-Elliott, M.M., Becker-Hapak, M., et al., 2021. Combining AFM13, a bispecific CD30/CD16 antibody, with cytokine-activated blood and cord blood-derived NK cells facilitates CAR-like responses against CD30(+) malignancies. Clin. Cancer Res. 27 (13), 3744–3756. PubMed PMID: 33986022. Pubmed Central PMCID: 8254785.

Kiessling, R., Klein, E., Wigzell, H., 1975. "Natural" killer cells in the mouse. I. Cytotoxic cells with specificity for mouse Moloney leukemia cells. Specificity and distribution according to genotype. Eur. J. Immunol. 5 (2), 112–117. PubMed PMID: 1234049.

Kim, S., Poursine-Laurent, J., Truscott, S.M., Lybarger, L., Song, Y.J., Yang, L., et al., 2005. Licensing of natural killer cells by host major histocompatibility complex class I molecules. Nature 436 (7051), 709–713. PubMed PMID: 16079848.

Kim, Y., Yoo, J.Y., Lee, T.J., Liu, J., Yu, J., Caligiuri, M.A., et al., 2018. Complex role of NK cells in regulation of oncolytic virus-bortezomib therapy. Proc. Natl. Acad. Sci. U. S. A. 115 (19), 4927–4932. PubMed PMID: 29686060. Pubmed Central PMCID: 5948955.

Kim, B.G., Malek, E., Choi, S.H., Ignatz-Hoover, J.J., Driscoll, J.J., 2021a. Novel therapies emerging in oncology to target the TGF-beta pathway. J. Hematol. Oncol. 14 (1), 55. PubMed PMID: 33823905. Pubmed Central PMCID: 8022551.

Kim, K.J., Moon, D., Kong, S.J., Lee, Y.S., Yoo, Y., Kim, S., et al., 2021b. Antitumor effects of IL-12 and GM-CSF co-expressed in an engineered oncolytic HSV-1. Gene Ther. 28 (3-4), 186–198. PubMed PMID: 33149278.

Kim, J., Kang, S., Kim, K.W., Heo, M.G., Park, D.I., Lee, J.H., et al., 2022. Nanoparticle delivery of recombinant IL-2 (BALLkine-2) achieves durable tumor control with less systemic adverse effects in cancer immunotherapy. Biomaterials 280, 121257. PubMed PMID: 34839122.

Klingemann, H., Boissel, L., Toneguzzo, F., 2016. Natural killer cells for immunotherapy—advantages of the NK-92 cell line over blood NK cells. Front. Immunol. 7, 91. PubMed PMID: 27014270. Pubmed Central PMCID: PMC4789404. Epub 2016/03/26.

Klose, C., Berchtold, S., Schmidt, M., Beil, J., Smirnow, I., Venturelli, S., et al., 2019. Biological treatment of pediatric sarcomas by combined virotherapy and NK cell therapy. BMC Cancer 19 (1), 1172. PubMed PMID: 31795974. Pubmed Central PMCID: 6889644.

Knorr, D.A., Ni, Z., Hermanson, D., Hexum, M.K., Bendzick, L., Cooper, L.J., et al., 2013. Clinical-scale derivation of natural killer cells from human pluripotent stem cells for cancer therapy. Stem Cells Transl. Med. 2 (4), 274–283. PubMed PMID: 23515118. Pubmed Central PMCID: PMC3659832. Epub 2013/03/22.

Kohrt, H.E., Thielens, A., Marabelle, A., Sagiv-Barfi, I., Sola, C., Chanuc, F., et al., 2014. Anti-KIR antibody enhancement of anti-lymphoma activity of natural killer cells as monotherapy and in combination with anti-CD20 antibodies. Blood 123 (5), 678–686. PubMed PMID: 24326534. Pubmed Central PMCID: 3907754.

Korde, N., Carlsten, M., Lee, M.J., Minter, A., Tan, E., Kwok, M., et al., 2014. A phase II trial of pan-KIR2D blockade with IPH2101 in smoldering multiple myeloma. Haematologica 99 (6), e81–e83. PubMed PMID: 24658821. Pubmed Central PMCID: 4040899.

Lahoz-Beneytez, J., Schaller, S., Macallan, D., Eissing, T., Niederalt, C., Asquith, B., 2017. Physiologically based simulations of deuterated glucose for quantifying cell turnover in humans. Front. Immunol. 8, 474. PubMed PMID: 28487698. Pubmed Central PMCID: 5403812.

Lakshman, A., Kumar, S.K., 2021. Chimeric antigen receptor T-cells, bispecific antibodies, and antibody-drug conjugates for multiple myeloma: an update. Am. J. Hematol. PubMed PMID: 34661922. Epub 2021/10/19.

Lapteva, N., Aldrich, M., Weksberg, D., Rollins, L., Goltsova, T., Chen, S.Y., et al., 2009. Targeting the intratumoral dendritic cells by the oncolytic adenoviral vaccine expressing RANTES elicits potent antitumor immunity. J. Immunother. 32 (2), 145–156. PubMed PMID: 19238013. Pubmed Central PMCID: 4146345.

Lemoine, J., Ruella, M., Houot, R., 2021. Born to survive: how cancer cells resist CAR T cell therapy. J. Hematol. Oncol. 14 (1), 199. PubMed PMID: 34809678. Pubmed Central PMCID: PMC8609883. Epub 2021/11/24.

Lenzi, R., Edwards, R., June, C., Seiden, M.V., Garcia, M.E., Rosenblum, M., et al., 2007. Phase II study of intraperitoneal recombinant interleukin-12 (rhIL-12) in patients with peritoneal carcinomatosis (residual disease <1 cm) associated with ovarian cancer or primary peritoneal carcinoma. J. Transl. Med. 5, 66. PubMed PMID: 18076766. Pubmed Central PMCID: 2248163.

Leonard, J.P., Sherman, M.L., Fisher, G.L., Buchanan, L.J., Larsen, G., Atkins, M.B., et al., 1997. Effects of single-dose interleukin-12 exposure on interleukin-12-associated toxicity and interferon-gamma production. Blood 90 (7), 2541–2548. PubMed PMID: 9326219.

Leung, E.Y.L., McNeish, I.A., 2021. Strategies to optimise oncolytic viral therapies: the role of natural killer cells. Viruses 13 (8). PubMed PMID: 34452316. Pubmed Central PMCID: 8402671.

Levy, E.R., Carlsten, M., Childs, R.W., 2016. mRNA transfection to improve NK cell homing to tumors. Methods Mol. Biol. 1441, 231–240. PubMed PMID: 27177670.

Levy, E., Reger, R., Segerberg, F., Lambert, M., Leijonhufvud, C., Baumer, Y., et al., 2019. Enhanced bone marrow homing of natural killer cells following mRNA transfection with gain-of-function variant CXCR4(R334X). Front. Immunol. 10, 1262. PubMed PMID: 31231387. Pubmed Central PMCID: 6560173.

Li, T., Yang, Y., Hua, X., Wang, G., Liu, W., Jia, C., et al., 2012. Hepatocellular carcinoma-associated fibroblasts trigger NK cell dysfunction via PGE2 and IDO. Cancer Lett. 318 (2), 154–161. PubMed PMID: 22182446.

Li, F., Sheng, Y., Hou, W., Sampath, P., Byrd, D., Thorne, S., et al., 2020. CCL5-armed oncolytic virus augments CCR5-engineered NK cell infiltration and antitumor efficiency. J. Immunotherapy Cancer 8 (1). PubMed PMID: 32098828. Pubmed Central PMCID: 7057442.

Li, H.K., Hsiao, C.W., Yang, S.H., Yang, H.P., Wu, T.S., Lee, C.Y., et al., 2021a. A novel off-the-shelf trastuzumab-armed NK cell therapy (ACE1702) using antibody-cell-conjugation technology. Cancer 31, 13(11). PubMed PMID: 34072864. Pubmed Central PMCID: 8199224.

Li, R., Zhang, J., Gilbert, S.M., Conejo-Garcia, J., Mule, J.J., 2021b. Using oncolytic viruses to ignite the tumour immune microenvironment in bladder cancer. Nat. Rev. Urol. 18 (9), 543–555. PubMed PMID: 34183833.

Lin, C., Ren, W., Luo, Y., Li, S., Chang, Y., Li, L., et al., 2020. Intratumoral delivery of a PD-1-blocking scFv encoded in oncolytic HSV-1 promotes antitumor immunity and synergizes with TIGIT blockade. Cancer Immunol. Res. 8 (5), 632–647. PubMed PMID: 32127389.

Liu, L.W., Nishikawa, T., Kaneda, Y., 2016. An RNA molecule derived from Sendai virus DI particles induces antitumor immunity and cancer cell-selective apoptosis. Mol. Ther. 24 (1), 135–145. PubMed PMID: 26548591. Pubmed Central PMCID: 4754554.

Liu, W., Wei, X., Li, L., Wu, X., Yan, J., Yang, H., et al., 2017a. CCR4 mediated chemotaxis of regulatory T cells suppress the activation of T cells and NK cells via TGF-beta pathway in human non-small cell lung cancer. Biochem. Biophys. Res. Commun. 488 (1), 196–203. PubMed PMID: 28487109.

Liu, Y., Cheng, Y., Xu, Y., Wang, Z., Du, X., Li, C., et al., 2017b. Increased expression of programmed cell death protein 1 on NK cells inhibits NK-cell-mediated anti-tumor function and indicates poor prognosis in digestive cancers. Oncogene 36 (44), 6143–6153. PubMed PMID: 28692048. Pubmed Central PMCID: 5671935.

Liu, E., Tong, Y., Dotti, G., Shaim, H., Savoldo, B., Mukherjee, M., et al., 2018. Cord blood NK cells engineered to express IL-15 and a CD19-targeted CAR show long-term persistence and potent antitumor activity. Leukemia 32 (2), 520–531. PubMed PMID: 28725044. Pubmed Central PMCID: PMC6063081. Epub 2017/07/21.

Liu, F., Huang, J., He, F., Ma, X., Fan, F., Meng, M., et al., 2020a. CD96, a new immune checkpoint, correlates with immune profile and clinical outcome of glioma. Sci. Rep. 10 (1), 10768. PubMed PMID: 32612110. Pubmed Central PMCID: 7330044.

Liu, E., Marin, D., Banerjee, P., Macapinlac, H.A., Thompson, P., Basar, R., et al., 2020b. Use of CAR-transduced natural killer cells in CD19-positive lymphoid tumors. N. Engl. J. Med. 382 (6), 545–553. PubMed PMID: 32023374. Pubmed Central PMCID: 7101242.

Liu, G., Zhang, Q., Yang, J., Li, X., Xian, L., Li, W., et al., 2021a. Increased TIGIT expressing NK cells with dysfunctional phenotype in AML patients correlated with poor prognosis. Cancer Immunol. Immunother. CII 15. PubMed PMID: 34129052.

Liu, S., Zhang, J., Fang, S., Zhang, Q., Zhu, G., Tian, Y., et al., 2021b. Macrophage polarization contributes to the efficacy of an oncolytic HSV-1 targeting human uveal melanoma in a murine xenograft model. Exp. Eye Res. 202, 108285. PubMed PMID: 33039456.

Lozano, E., Mena, M.P., Diaz, T., Martin-Antonio, B., Leon, S., Rodriguez-Lobato, L.G., et al., 2020. Nectin-2 expression on malignant plasma cells is associated with better response to TIGIT blockade in multiple myeloma. Clin. Cancer Res. 26 (17), 4688–4698. PubMed PMID: 32513837.

Lu, C., Guo, C., Chen, H., Zhang, H., Zhi, L., Lv, T., et al., 2020a. A novel chimeric PD1-NKG2D-41BB receptor enhances antitumor activity of NK92 cells against human lung cancer H1299 cells by triggering pyroptosis. Mol. Immunol. 122, 200–206. PubMed PMID: 32388482. Epub 2020/05/11.

Lu, Y., Xue, J., Deng, T., Zhou, X., Yu, K., Deng, L., et al., 2020b. Safety and feasibility of CRISPR-edited T cells in patients with refractory non-small-cell lung cancer. Nat. Med. 26 (5), 732–740. PubMed PMID: 32341578.

Luanpitpong, S., Poohadsuan, J., Klaihmon, P., Issaragrisil, S., 2021. Selective cytotoxicity of single and dual anti-CD19 and anti-CD138 chimeric antigen receptor-natural killer cells against hematologic malignancies. J. Immunol. Res. 2021, 5562630. PubMed PMID: 34337077. Pubmed Central PMCID: PMC8289607. Epub 2021/08/03.

Luevano, M., Daryouzeh, M., Alnabhan, R., Querol, S., Khakoo, S., Madrigal, A., et al., 2012. The unique profile of cord blood natural killer cells balances incomplete maturation and effective killing function upon activation. Hum. Immunol. 73 (3), 248–257. PubMed PMID: 22234167. Epub 2012/01/12.

Ma, L., Gai, J., Qiao, P., Li, Y., Li, X., Zhu, M., et al., 2020. A novel bispecific nanobody with PD-L1/TIGIT dual immune checkpoint blockade. Biochem. Biophys. Res. Commun. 531 (2), 144–151. PubMed PMID: 32782142.

Ma, R., Lu, T., Li, Z., Teng, K.Y., Mansour, A.G., Yu, M., et al., 2021. An oncolytic virus expressing IL15/IL15Ralpha combined with off-the-shelf EGFR-CAR NK cells targets glioblastoma. Cancer Res. 81 (13), 3635–3648. PubMed PMID: 34006525. Pubmed Central PMCID: 8562586.

Mancusi, A., Alvarez, M., Piccinelli, S., Velardi, A., Pierini, A., 2019. TNFR2 signaling modulates immunity after allogeneic hematopoietic cell transplantation. Cytokine Growth Factor Rev. 47, 54–61. PubMed PMID: 31122819.

Mao, L., Xiao, Y., Yang, Q.C., Yang, S.C., Yang, L.L., Sun, Z.J., 2021. TIGIT/CD155 blockade enhances anti-PD-L1 therapy in head and neck squamous cell carcinoma by targeting myeloid-derived suppressor cells. Oral Oncol. 121, 105472. PubMed PMID: 34333450.

Marcoe, J.P., Lim, J.R., Schaubert, K.L., Fodil-Cornu, N., Matka, M., McCubbrey, A.L., et al., 2012. TGF-beta is responsible for NK cell immaturity during ontogeny and increased susceptibility to infection during mouse infancy. Nat. Immunol. 13 (9), 843–850. PubMed PMID: 22863752. Pubmed Central PMCID: 3426626.

Margolin, K., Morishima, C., Velcheti, V., Miller, J.S., Lee, S.M., Silk, A.W., et al., 2018. Phase I trial of ALT-803, a novel recombinant IL15 complex, in patients with advanced solid tumors. Clin. Cancer Res. 24 (22), 5552–5561. PubMed PMID: 30045932. Pubmed Central PMCID: 6239933.

Marofi, F., Al-Awad, A.S., Sulaiman Rahman, H., Markov, A., Abdelbasset, W.K., Ivanovna Enina, Y., et al., 2021. CAR-NK cell: a new paradigm in tumor immunotherapy. Front. Oncol. 11, 673276. PubMed PMID: 34178661. Pubmed Central PMCID: 8223062.

Marotel, M., Hasim, M.S., Hagerman, A., Ardolino, M., 2020. The two-faces of NK cells in oncolytic virotherapy. Cytokine Growth Factor Rev. 56, 59–68. PubMed PMID: 32586674.

Martin, C.J., Datta, A., Littlefield, C., Kalra, A., Chapron, C., Wawersik, S., et al., 2020. Selective inhibition of TGFbeta1 activation overcomes primary resistance to checkpoint blockade therapy by altering tumor immune landscape. Sci. Transl. Med. 12 (536). PubMed PMID: 32213632.

Masu, T., Atsukawa, M., Nakatsuka, K., Shimizu, M., Miura, D., Arai, T., et al., 2018. Anti-CD137 monoclonal antibody enhances trastuzumab-induced, natural killer cell-mediated cytotoxicity against pancreatic cancer cell lines with low human epidermal growth factor-like receptor 2 expression. PLoS One 13 (12), e0200664. PubMed PMID: 30596643. Pubmed Central PMCID: 6312288.

McMichael, E.L., Benner, B., Atwal, L.S., Courtney, N.B., Mo, X., Davis, M.E., et al., 2019. A Phase I/II trial of Cetuximab in combination with interleukin-12 administered to patients with unresectable primary or recurrent head and neck squamous cell carcinoma. Clin. Cancer Res. 25 (16), 4955–4965. PubMed PMID: 31142501. Pubmed Central PMCID: 6697573.

Mealiea, D., McCart, J.A., 2021. Cutting both ways: the innate immune response to oncolytic virotherapy. Cancer Gene Ther. PubMed PMID: 34453122.

Medina-Echeverz, J., Hinterberger, M., Testori, M., Geiger, M., Giessel, R., Bathke, B., et al., 2019. Synergistic cancer immunotherapy combines MVA-CD40L induced innate and adaptive immunity with tumor targeting antibodies. Nat. Commun. 10 (1), 5041. PubMed PMID: 31695037. Pubmed Central PMCID: 6834557.

Meinhardt, K., Kroeger, I., Bauer, R., Ganss, F., Ovsiy, I., Rothamer, J., et al., 2015. Identification and characterization of the specific murine NK cell subset supporting graft-versus-leukemia- and reducing graft-versus-host-effects. Oncoimmunology. 4 (1), e981483. PubMed PMID: 25949862. Pubmed Central PMCID: PMC4368119. Epub 2015/05/08.

Melaiu, O., Lucarini, V., Cifaldi, L., Fruci, D., 2019. Influence of the tumor microenvironment on NK cell function in solid tumors. Front. Immunol. 10, 3038. PubMed PMID: 32038612. Pubmed Central PMCID: 6985149.

Melero, I., Shuford, W.W., Newby, S.A., Aruffo, A., Ledbetter, J.A., Hellstrom, K.E., et al., 1997. Monoclonal antibodies against the 4-1BB T-cell activation molecule eradicate established tumors. Nat. Med. 3 (6), 682–685. PubMed PMID: 9176498.

Melero, I., Johnston, J.V., Shufford, W.W., Mittler, R.S., Chen, L., 1998. NK1.1 cells express 4-1BB (CDw137) costimulatory molecule and are required for tumor immunity elicited by anti-4-1BB monoclonal antibodies. Cell. Immunol. 190 (2), 167–172. PubMed PMID: 9878117.

Melero, I., Castanon, E., Alvarez, M., Champiat, S., Marabelle, A., 2021. Intratumoural administration and tumour tissue targeting of cancer immunotherapies. Nat. Rev. Clin. Oncol. 18 (9), 558–576. PubMed PMID: 34006998. Pubmed Central PMCID: 8130796.

Merino, A., Zhang, B., Dougherty, P., Luo, X., Wang, J., Blazar, B.R., et al., 2019. Chronic stimulation drives human NK cell dysfunction and epigenetic reprograming. J. Clin.

Invest. 129 (9), 3770–3785. PubMed PMID: 31211698. Pubmed Central PMCID: 6715389.

Mettu, N.B., Ulahannan, S.V., Bendell, J.C., Garrido-Laguna, I., Strickler, J.H., Moore, K.N., et al., 2021. A phase 1a/b OpenLabel, DoseEscalation Study of Etigilimab alone or in combination with nivolumab in patients with locally advanced or metastatic solid tumors. Clin. Cancer Res. 29. PubMed PMID: 34844977.

Miller, C.G., Fraser, N.W., 2003. Requirement of an integrated immune response for successful neuroattenuated HSV-1 therapy in an intracranial metastatic melanoma model. Mol. Ther. 7 (6), 741–747. PubMed PMID: 12788647. Pubmed Central PMCID: 2661757.

Miller, J.S., Tessmer-Tuck, J., Pierson, B.A., Weisdorf, D., McGlave, P., Blazar, B.R., et al., 1997. Low dose subcutaneous interleukin-2 after autologous transplantation generates sustained in vivo natural killer cell activity. Biol. Blood Marrow Transplant. 3 (1), 34–44. PubMed PMID: 9209739.

Miller, J.S., Soignier, Y., Panoskaltsis-Mortari, A., McNearney, S.A., Yun, G.H., Fautsch, S.K., et al., 2005. Successful adoptive transfer and in vivo expansion of human haploidentical NK cells in patients with cancer. Blood 105 (8), 3051–3057. PubMed PMID: 15632206.

Mittal, D., Lepletier, A., Madore, J., Aguilera, A.R., Stannard, K., Blake, S.J., et al., 2019. CD96 is an immune checkpoint that regulates CD8(+) T-cell antitumor function. Cancer Immunol. Res. 7 (4), 559–571. PubMed PMID: 30894377. Pubmed Central PMCID: 6445751.

Miyamoto, S., Inoue, H., Nakamura, T., Yamada, M., Sakamoto, C., Urata, Y., et al., 2012. Coxsackievirus B3 is an oncolytic virus with immunostimulatory properties that is active against lung adenocarcinoma. Cancer Res. 72 (10), 2609–2621. PubMed PMID: 22461509.

Morimoto, T., Nakazawa, T., Matsuda, R., Nishimura, F., Nakamura, M., Yamada, S., et al., 2021. CRISPR-Cas9-mediated TIM3 knockout in human natural killer cells enhances growth inhibitory effects on human glioma cells. Int. J. Mol. Sci. 22 (7). PubMed PMID: 33800561. Pubmed Central PMCID: 8036491.

Morris, J.C., Tan, A.R., Olencki, T.E., Shapiro, G.I., Dezube, B.J., Reiss, M., et al., 2014. Phase I study of GC1008 (fresolimumab): a human anti-transforming growth factor-beta (TGFbeta) monoclonal antibody in patients with advanced malignant melanoma or renal cell carcinoma. PLoS One 9 (3), e90353. PubMed PMID: 24618589. Pubmed Central PMCID: 3949712.

Mortier, E., Quemener, A., Vusio, P., Lorenzen, I., Boublik, Y., Grotzinger, J., et al., 2006. Soluble interleukin-15 receptor alpha (IL-15R alpha)-sushi as a selective and potent agonist of IL-15 action through IL-15R beta/gamma. Hyperagonist IL-15 x IL-15R alpha fusion proteins. J. Biol. Chem. 281 (3), 1612–1619. PubMed PMID: 16284400.

Motzer, R.J., Rakhit, A., Schwartz, L.H., Olencki, T., Malone, T.M., Sandstrom, K., et al., 1998. Phase I trial of subcutaneous recombinant human interleukin-12 in patients with advanced renal cell carcinoma. Clin. Cancer Res. 4 (5), 1183–1191. PubMed PMID: 9607576.

Muller, T., Uherek, C., Maki, G., Chow, K.U., Schimpf, A., Klingemann, H.G., et al., 2008. Expression of a CD20-specific chimeric antigen receptor enhances cytotoxic activity of NK cells and overcomes NK-resistance of lymphoma and leukemia cells. Cancer Immunol. Immunother. 57 (3), 411–423. PubMed PMID: 17717662. Epub 2007/08/25.

Myers, J.A., Miller, J.S., 2021. Exploring the NK cell platform for cancer immunotherapy. Nat. Rev. Clin. Oncol. 18 (2), 85–100. PubMed PMID: 32934330. Pubmed Central PMCID: 8316981.

Naing, A., Papadopoulos, K.P., Autio, K.A., Ott, P.A., Patel, M.R., Wong, D.J., et al., 2016. Safety, antitumor activity, and immune activation of pegylated recombinant human interleukin-10 (AM0010) in patients with advanced solid tumors. J. Clin. Oncol. 34 (29), 3562–3569. PubMed PMID: 27528724. Pubmed Central PMCID: 5657013.

Naing, A., Wong, D.J., Infante, J.R., Korn, W.M., Aljumaily, R., Papadopoulos, K.P., et al., 2019. Pegilodecakin combined with pembrolizumab or nivolumab for patients with advanced solid tumours (IVY): a multicentre, multicohort, open-label, phase 1b trial. Lancet Oncol. 20 (11), 1544–1555. PubMed PMID: 31563517. Pubmed Central PMCID: 8436252.

Neo, S.Y., Yang, Y., Record, J., Ma, R., Chen, X., Chen, Z., et al., 2020. CD73 immune checkpoint defines regulatory NK cells within the tumor microenvironment. J. Clin. Invest. 130 (3), 1185–1198. PubMed PMID: 31770109. Pubmed Central PMCID: 7269592.

Ng, Y.Y., Du, Z., Zhang, X., Chng, W.J., Wang, S., 2021. CXCR4 and anti-BCMA CAR co-modified natural killer cells suppress multiple myeloma progression in a xenograft mouse model. Cancer Gene Ther. PubMed PMID: 34471234. Epub 2021/09/03.

Nguyen, L.T., Saibil, S.D., Sotov, V., Le, M.X., Khoja, L., Ghazarian, D., et al., 2019. Phase II clinical trial of adoptive cell therapy for patients with metastatic melanoma with autologous tumor-infiltrating lymphocytes and low-dose interleukin-2. Cancer Immunol. Immunother. CII 68 (5), 773–785. PubMed PMID: 30747243.

Ni, G., Zhang, L., Yang, X., Li, H., Ma, B., Walton, S., et al., 2020. Targeting interleukin-10 signalling for cancer immunotherapy, a promising and complicated task. Human Vaccin. Immunother. 16 (10), 2328–2332. PubMed PMID: 32159421. Pubmed Central PMCID: 7644214.

Nijhof, I.S., Lammerts van Bueren, J.J., van Kessel, B., Andre, P., Morel, Y., Lokhorst, H.M., et al., 2015. Daratumumab-mediated lysis of primary multiple myeloma cells is enhanced in combination with the human anti-KIR antibody IPH2102 and lenalidomide. Haematologica 100 (2), 263–268. PubMed PMID: 25510242. Pubmed Central PMCID: 4803142.

Niu, Z., Bai, F., Sun, T., Tian, H., Yu, D., Yin, J., et al., 2015. Recombinant Newcastle disease virus expressing IL15 demonstrates promising antitumor efficiency in melanoma model. Technol. Cancer Res. Treat. 14 (5), 607–615. PubMed PMID: 24645750.

Niu, J., Maurice-Dror, C., Lee, D.H., Kim, D.W., Nagrial, A., Voskoboynik, M., et al., 2021. First-in-human phase 1 study of the anti-TIGIT antibody vibostolimab as monotherapy or with pembrolizumab for advanced solid tumors, including non-small-cell lung cancer. Ann. Oncol. PubMed PMID: 34800678.

Nowakowska, P., Romanski, A., Miller, N., Odendahl, M., Bonig, H., Zhang, C., et al., 2018. Clinical grade manufacturing of genetically modified, CAR-expressing NK-92 cells for the treatment of ErbB2-positive malignancies. Cancer Immunol. Immunother. 67 (1), 25–38. PubMed PMID: 28879551. Epub 2017/09/08.

Ochoa, M.C., Fioravanti, J., Rodriguez, I., Hervas-Stubbs, S., Azpilikueta, A., Mazzolini, G., et al., 2013. Antitumor immunotherapeutic and toxic properties of an HDL-conjugated chimeric IL-15 fusion protein. Cancer Res. 73 (1), 139–149. PubMed PMID: 23149919.

Ochoa, M.C., Minute, L., Lopez, A., Perez-Ruiz, E., Gomar, C., Vasquez, M., et al., 2018. Enhancement of antibody-dependent cellular cytotoxicity of cetuximab by a chimeric protein encompassing interleukin-15. Oncoimmunology 7 (2), e1393579. PubMed PMID: 29308327. Pubmed Central PMCID: 5749662.

Ochoa, M.C., Perez-Ruiz, E., Minute, L., Onate, C., Perez, G., Rodriguez, I., et al., 2019. Daratumumab in combination with urelumab to potentiate anti-myeloma activity in lymphocyte-deficient mice reconstituted with human NK cells. Oncoimmunology 8 (7), 1599636. PubMed PMID: 31143521. Pubmed Central PMCID: 6527281.

Oelsner, S., Friede, M.E., Zhang, C., Wagner, J., Badura, S., Bader, P., et al., 2017. Continuously expanding CAR NK-92 cells display selective cytotoxicity against B-cell leukemia and lymphoma. Cytotherapy 19 (2), 235–249. PubMed PMID: 27887866. Epub 2016/11/27.

Ogbomo, H., Michaelis, M., Geiler, J., van Rikxoort, M., Muster, T., Egorov, A., et al., 2010. Tumor cells infected with oncolytic influenza A virus prime natural killer cells for lysis of resistant tumor cells. Med. Microbiol. Immunol. 199 (2), 93–101. PubMed PMID: 20012989.

Olson, J.A., Leveson-Gower, D.B., Gill, S., Baker, J., Beilhack, A., Negrin, R.S., 2010. NK cells mediate reduction of GVHD by inhibiting activated, alloreactive T cells while retaining GVT effects. Blood 115 (21), 4293–4301. PubMed PMID: 20233969. Pubmed Central PMCID: PMC2879101. Epub 2010/03/18.

Ong, H.T., Timm, M.M., Greipp, P.R., Witzig, T.E., Dispenzieri, A., Russell, S.J., et al., 2006. Oncolytic measles virus targets high CD46 expression on multiple myeloma cells. Exp. Hematol. 34 (6), 713–720. PubMed PMID: 16728275.

Otano, I., Azpilikueta, A., Glez-Vaz, J., Alvarez, M., Medina-Echeverz, J., Cortes-Dominguez, I., et al., 2021. CD137 (4-1BB) costimulation of CD8(+) T cells is more potent when provided in cis than in trans with respect to CD3-TCR stimulation. Nat. Commun. 12 (1), 7296. PubMed PMID: 34911975. Pubmed Central PMCID: 8674279.

Pardi, N., Hogan, M.J., Porter, F.W., Weissman, D., 2018. mRNA vaccines—a new era in vaccinology. Nat. Rev. Drug Discov. 17 (4), 261–279. PubMed PMID: 29326426. Pubmed Central PMCID: 5906799.

Parihar, R., Dierksheide, J., Hu, Y., Carson, W.E., 2002. IL-12 enhances the natural killer cell cytokine response to Ab-coated tumor cells. J. Clin. Invest. 110 (7), 983–992. PubMed PMID: 12370276. Pubmed Central PMCID: 151155.

Pastor, F., Berraondo, P., Etxeberria, I., Frederick, J., Sahin, U., Gilboa, E., et al., 2018. An RNA toolbox for cancer immunotherapy. Nat. Rev. Drug Discov. 17 (10), 751–767. PubMed PMID: 30190565.

Payne, R., Glenn, L., Hoen, H., Richards, B., Smith 2nd, J.W., Lufkin, R., et al., 2014. Durable responses and reversible toxicity of high-dose interleukin-2 treatment of melanoma and renal cancer in a Community Hospital Biotherapy Program. J. Immunotherapy Cancer 2, 13. PubMed PMID: 24855563. Pubmed Central PMCID: 4030280.

Paz-Ares, L., Kim, T.M., Vicente, D., Felip, E., Lee, D.H., Lee, K.H., et al., 2020. Bintrafusp Alfa, a bifunctional fusion protein targeting TGF-beta and PD-L1, in second-line treatment of patients with NSCLC: results from an expansion cohort of a Phase 1 trial. J. Thorac. Oncol. 15 (7), 1210–1222. PubMed PMID: 32173464. Pubmed Central PMCID: 8210474.

Phung, S.K., Miller, J.S., Felices, M., 2021. Bi-specific and tri-specific NK cell engagers: the new avenue of targeted NK cell immunotherapy. Mol. Diagn. Ther. 25 (5), 577–592. PubMed PMID: 34327614.

Pierini, A., Alvarez, M., Negrin, R.S., 2016. NK cell and CD4+FoxP3+ regulatory T cell based therapies for hematopoietic stem cell engraftment. Stem Cells Int. 2016, 9025835. PubMed PMID: 26880996. Pubmed Central PMCID: 4736409.

Plaks, V., Rossi, J.M., Chou, J., Wang, L., Poddar, S., Han, G., et al., 2021. CD19 target evasion as a mechanism of relapse in large B-cell lymphoma treated with axicabtagene ciloleucel. Blood 138 (12), 1081–1085. PubMed PMID: 34041526. Pubmed Central PMCID: PMC8462361. Epub 2021/05/28.

Ptacin, J.L., Caffaro, C.E., Ma, L., San Jose Gall, K.M., Aerni, H.R., Acuff, N.V., et al., 2021. An engineered IL-2 reprogrammed for anti-tumor therapy using a semi-synthetic organism. Nat. Commun. 12 (1), 4785. PubMed PMID: 34373459. Pubmed Central PMCID: 8352909.

Qiao, J., Liu, Z., Dong, C., Luan, Y., Zhang, A., Moore, C., et al., 2019. Targeting tumors with IL-10 prevents dendritic cell-mediated CD8(+) T cell apoptosis. Cancer Cell 35 (6), 901–915. e4. PubMed PMID: 31185213.

Quetglas, J.I., Labiano, S., Aznar, M.A., Bolanos, E., Azpilikueta, A., Rodriguez, I., et al., 2015. Virotherapy with a Semliki Forest Virus-based vector encoding IL12 synergizes with PD-1/PD-L1 blockade. Cancer Immunol. Res. 3 (5), 449–454. PubMed PMID: 25691326.

Rallis, K.S., Corrigan, A.E., Dadah, H., George, A.M., Keshwara, S.M., Sideris, M., et al., 2021. Cytokine-based cancer immunotherapy: challenges and opportunities for IL-10. Anticancer Res. 41 (7), 3247–3252. PubMed PMID: 34230118.

Ramelyte, E., Tastanova, A., Balazs, Z., Ignatova, D., Turko, P., Menzel, U., et al., 2021. Oncolytic virotherapy-mediated anti-tumor response: a single-cell perspective. Cancer Cell 39 (3), 394–406 e4. PubMed PMID: 33482123.

Raulet, D.H., 2006. Missing self recognition and self tolerance of natural killer (NK) cells. Semin. Immunol. 18 (3), 145–150. PubMed PMID: 16740393.

Ravi, R., Noonan, K.A., Pham, V., Bedi, R., Zhavoronkov, A., Ozerov, I.V., et al., 2018. Bifunctional immune checkpoint-targeted antibody-ligand traps that simultaneously disable TGFbeta enhance the efficacy of cancer immunotherapy. Nat. Commun. 9 (1), 741. PubMed PMID: 29467463. Pubmed Central PMCID: 5821872.

Robbins, G.M., Wang, M., Pomeroy, E.J., Moriarity, B.S., 2021. Nonviral genome engineering of natural killer cells. Stem Cell Res Ther 12 (1), 350. PubMed PMID: 34134774. Pubmed Central PMCID: 8207670.

Romagne, F., Andre, P., Spee, P., Zahn, S., Anfossi, N., Gauthier, L., et al., 2009. Preclinical characterization of 1-7F9, a novel human anti-KIR receptor therapeutic antibody that augments natural killer-mediated killing of tumor cells. Blood 114 (13), 2667–2677. PubMed PMID: 19553639. Pubmed Central PMCID: 2756126.

Roman Aguilera, A., Lutzky, V.P., Mittal, D., Li, X.Y., Stannard, K., Takeda, K., et al., 2018. CD96 targeted antibodies need not block CD96-CD155 interactions to promote NK cell anti-metastatic activity. Oncoimmunology 7 (5), e1424677. PubMed PMID: 29721390. Pubmed Central PMCID: 5927540.

Romanski, A., Uherek, C., Bug, G., Seifried, E., Klingemann, H., Wels, W.S., et al., 2016. CD19-CAR engineered NK-92 cells are sufficient to overcome NK cell resistance in B-cell malignancies. J. Cell. Mol. Med. 20 (7), 1287–1294. PubMed PMID: 27008316. Pubmed Central PMCID: PMC4929308. Epub 2016/03/24.

Romee, R., Rosario, M., Berrien-Elliott, M.M., Wagner, J.A., Jewell, B.A., Schappe, T., et al., 2016. Cytokine-induced memory-like natural killer cells exhibit enhanced responses against myeloid leukemia. Sci. Transl. Med. 8 (357). 357ra123 PubMed PMID: 27655849. Pubmed Central PMCID: 5436500.

Rosenberg, S.A., Lotze, M.T., Muul, L.M., Leitman, S., Chang, A.E., Vetto, J.T., et al., 1986. A new approach to the therapy of cancer based on the systemic administration of autologous lymphokine-activated killer cells and recombinant interleukin-2. Surgery 100 (2), 262–272. PubMed PMID: 3526604.

Rothe, A., Sasse, S., Topp, M.S., Eichenauer, D.A., Hummel, H., Reiners, K.S., et al., 2015. A phase 1 study of the bispecific anti-CD30/CD16A antibody construct AFM13 in patients with relapsed or refractory Hodgkin lymphoma. Blood 125 (26), 4024–4031. PubMed PMID: 25887777. Pubmed Central PMCID: 4528081.

Ruggeri, L., Capanni, M., Urbani, E., Perruccio, K., Shlomchik, W.D., Tosti, A., et al., 2002. Effectiveness of donor natural killer cell alloreactivity in mismatched hematopoietic transplants. Science 295 (5562), 2097–2100. PubMed PMID: 11896281. Epub 2002/03/16.

Safarzadeh Kozani, P., Safarzadeh Kozani, P., Rahbarizadeh, F., 2021. Optimizing the clinical impact of CAR-T cell therapy in B-cell acute lymphoblastic leukemia: looking back while moving forward. Front. Immunol. 12, 765097. PubMed PMID: 34777381. Pubmed Central PMCID: PMC8581403. Epub 2021/11/16.

Sahin, U., Kariko, K., Tureci, O., 2014. mRNA-based therapeutics—developing a new class of drugs. Nat. Rev. Drug Discov. 13 (10), 759–780. PubMed PMID: 25233993.

Sahin, D., Arenas-Ramirez, N., Rath, M., Karakus, U., Humbelin, M., van Gogh, M., et al., 2020. An IL-2-grafted antibody immunotherapy with potent efficacy against metastatic cancer. Nat. Commun. 11 (1), 6440. PubMed PMID: 33353953. Pubmed Central PMCID: 7755894.

Sanchez-Correa, B., Lopez-Sejas, N., Duran, E., Labella, F., Alonso, C., Solana, R., et al., 2019. Modulation of NK cells with checkpoint inhibitors in the context of cancer immunotherapy. Cancer Immunol. Immunother. CII 68 (5), 861–870. PubMed PMID: 30953117.

Sanmamed, M.F., Pastor, F., Rodriguez, A., Perez-Gracia, J.L., Rodriguez-Ruiz, M.E., Jure-Kunkel, M., et al., 2015. Agonists of co-stimulation in cancer immunotherapy directed against CD137, OX40, GITR, CD27, CD28, and ICOS. Semin. Oncol. 42 (4), 640–655. PubMed PMID: 26320067.

Sarhan, D., Brandt, L., Felices, M., Guldevall, K., Lenvik, T., Hinderlie, P., et al., 2018. 161533 TriKE stimulates NK-cell function to overcome myeloid-derived suppressor cells in MDS. Blood Adv. 2 (12), 1459–1469. PubMed PMID: 29941459. Pubmed Central PMCID: 6020813.

Schlake, T., Thess, A., Thran, M., Jordan, I., 2019. mRNA as novel technology for passive immunotherapy. Cell. Mol. Life Sci. CMLS 76 (2), 301–328. PubMed PMID: 30334070. Pubmed Central PMCID: 6339677.

Schmohl, J.U., Felices, M., Taras, E., Miller, J.S., Vallera, D.A., 2016a. Enhanced ADCC and NK cell activation of an anticarcinoma bispecific antibody by genetic insertion of a modified IL-15 cross-linker. Mol. Ther. 24 (7), 1312–1322. PubMed PMID: 27157665. Pubmed Central PMCID: 5088765.

Schmohl, J.U., Gleason, M.K., Dougherty, P.R., Miller, J.S., Vallera, D.A., 2016b. Heterodimeric bispecific single chain variable fragments (scFv) killer engagers (BiKEs) enhance NK-cell activity against CD133+ colorectal cancer cells. Target. Oncol. 11 (3), 353–361. PubMed PMID: 26566946. Pubmed Central PMCID: 4873478.

Schmohl, J.U., Felices, M., Oh, F., Lenvik, A.J., Lebeau, A.M., Panyam, J., et al., 2017. Engineering of anti-CD133 Trispecific molecule capable of inducing NK expansion and driving antibody-dependent cell-mediated cytotoxicity. Cancer Res. Treat. 49 (4), 1140–1152. PubMed PMID: 28231426. Pubmed Central PMCID: 5654165.

Schwaiger, T., Knittler, M.R., Grund, C., Roemer-Oberdoerfer, A., Kapp, J.F., Lerch, M.M., et al., 2017. Newcastle disease virus mediates pancreatic tumor rejection via NK cell activation and prevents cancer relapse by prompting adaptive immunity. Int. J. Cancer 141 (12), 2505–2516. PubMed PMID: 28857157.

Senekal, N.S., Mahasa, K.J., Eladdadi, A., de Pillis, L., Ouifki, R., 2021. Natural killer cells recruitment in oncolytic virotherapy: a mathematical model. Bull. Math. Biol. 83 (7), 75. PubMed PMID: 34008149.

Shah, N.N., Baird, K., Delbrook, C.P., Fleisher, T.A., Kohler, M.E., Rampertaap, S., et al., 2015. Acute GVHD in patients receiving IL-15/4-1BBL activated NK cells following T-cell-depleted stem cell transplantation. Blood 125 (5), 784–792. PubMed PMID: 25452614. Pubmed Central PMCID: PMC4311226. Epub 2014/12/03.

Shirasuna, K., Koelsch, G., Seidel-Dugan, C., Salmeron, A., Steiner, P., Winston, W.M., et al., 2021. Characterization of ASP8374, a fully-human, antagonistic anti-TIGIT monoclonal antibody. Cancer Treat. Res. Commun. 28, 100433. PubMed PMID: 34273876.

Simonetta, F., Alvarez, M., Negrin, R.S., 2017. Natural killer cells in graft-versus-host-disease after allogeneic hematopoietic cell transplantation. Front. Immunol. 8, 465. PubMed PMID: 28487696. Pubmed Central PMCID: 5403889.

Song, X., Si, Q., Qi, R., Liu, W., Li, M., Guo, M., et al., 2021. Indoleamine 2,3-dioxygenase 1: a promising therapeutic target in malignant tumor. Front. Immunol. 12, 800630. PubMed PMID: 35003126. Pubmed Central PMCID: 8733291.

Speck, T, Heidbuechel, JPW, Veinalde, R, Jaeger, D, von Kalle, C, Ball, CR, et al., 2018. Targeted BiTE expression by an oncolytic vector augments therapeutic efficacy against solid tumors. Clin. Cancer Res. 24 (9), 2128–2137. PubMed PMID: 29437789.

Spits, H., Bernink, J.H., Lanier, L., 2016. NK cells and type 1 innate lymphoid cells: partners in host defense. Nat. Immunol. 17 (7), 758–764. PubMed PMID: 27328005.

Srivastava, R.M., Trivedi, S., Concha-Benavente, F., Gibson, S.P., Reeder, C., Ferrone, S., et al., 2017. CD137 stimulation enhances cetuximab-induced natural killer: dendritic cell priming of antitumor t-cell immunity in patients with head and neck cancer. Clin. Cancer Res. 23 (3), 707–716. PubMed PMID: 27496866. Pubmed Central PMCID: 5290200.

Stephenson, K.B., Barra, N.G., Davies, E., Ashkar, A.A., Lichty, B.D., 2012. Expressing human interleukin-15 from oncolytic vesicular stomatitis virus improves survival in a murine metastatic colon adenocarcinoma model through the enhancement of anti-tumor immunity. Cancer Gene Ther. 19 (4), 238–246. PubMed PMID: 22158521.

Strauss, J., Heery, C.R., Kim, J.W., Jochems, C., Donahue, R.N., Montgomery, A.S., et al., 2019. First-in-human Phase I trial of a tumor-targeted cytokine (NHS-IL12) in subjects with metastatic solid tumors. Clin. Cancer Res. 25 (1), 99–109. PubMed PMID: 30131389. Pubmed Central PMCID: 6320276.

Sun, K., Alvarez, M., Ames, E., Barao, I., Chen, M., Longo, D.L., et al., 2012. Mouse NK cell-mediated rejection of bone marrow allografts exhibits patterns consistent with Ly49 subset licensing. Blood 119 (6), 1590–1598. PubMed PMID: 22184406. Pubmed Central PMCID: 3286220.

Sun, Y., Wang, S., Yang, H., Wu, J., Li, S., Qiao, G., et al., 2019. Impact of synchronized anti-PD-1 with Ad-CEA vaccination on inhibition of colon cancer growth. Immunotherapy 11 (11), 953–966. PubMed PMID: 31192764.

Sun, R., Xiong, Y., Liu, H., Gao, C., Su, L., Weng, J., et al., 2020. Tumor-associated neutrophils suppress antitumor immunity of NK cells through the PD-L1/PD-1 axis. Transl. Oncol. 13 (10), 100825. PubMed PMID: 32698059. Pubmed Central PMCID: 7372151.

Tang, X., Yang, L., Li, Z., Nalin, A.P., Dai, H., Xu, T., et al., 2018. First-in-man clinical trial of CAR NK-92 cells: safety test of CD33-CAR NK-92 cells in patients with relapsed and refractory acute myeloid leukemia. Am. J. Cancer Res. 8 (6), 1083–1089. PubMed PMID: 30034945. Pubmed Central PMCID: PMC6048396. Epub 2018/07/24.

Tannir, N.M., Papadopoulos, K.P., Wong, D.J., Aljumaily, R., Hung, A., Afable, M., et al., 2021. Pegilodecakin as monotherapy or in combination with anti-PD-1 or tyrosine kinase inhibitor in heavily pretreated patients with advanced renal cell carcinoma: final results of cohorts A, G, H and I of IVY Phase I study. Int. J. Cancer 149 (2), 403–408. PubMed PMID: 33709428. Pubmed Central PMCID: 8251721.

Tassev, D.V., Cheng, M., Cheung, N.K., 2012. Retargeting NK92 cells using an HLA-A2-restricted, EBNA3C-specific chimeric antigen receptor. Cancer Gene Ther. 19 (2), 84–100. PubMed PMID: 21979579. Epub 2011/10/08.

Tauriello, D.V.F., Sancho, E., Batlle, E., 2022. Overcoming TGFbeta-mediated immune evasion in cancer. Nat. Rev. Cancer 22 (1), 25–44. PubMed PMID: 34671117.

Therapeutics, F., 2021. Fate Therapeutics Announces Positive Interim Clinical Data from its FT596 and FT516 Off-the-Shelf, iPSC-Derived NK Cell Programs for B-Cell Lymphoma [Updated Aug 19 2021]. Available from: https://ir.fatetherapeutics.com/news-releases/news-release-details/fate-therapeutics-announces-positive-interim-clinical-data-its.

Thomas, S., Kuncheria, L., Roulstone, V., Kyula, J.N., Mansfield, D., Bommareddy, P.K., et al., 2019. Development of a new fusion-enhanced oncolytic immunotherapy platform based on herpes simplex virus type 1. J. Immunotherapy Cancer 7 (1), 214. PubMed PMID: 31399043. Pubmed Central PMCID: 6689178.

Tian, C., Liu, J., Zhou, H., Li, J., Sun, C., Zhu, W., et al., 2021. Enhanced anti-tumor response elicited by a novel oncolytic HSV-1 engineered with an anti-PD-1 antibody. Cancer Lett. 518, 49–58. PubMed PMID: 34139284.

Tinker, A.V., Hirte, H.W., Provencher, D., Butler, M., Ritter, H., Tu, D., et al., 2019. Dose-ranging and cohort-expansion study of monalizumab (IPH2201) in patients with advanced gynecologic malignancies: a trial of the Canadian Cancer Trials Group (CCTG): IND221. Clin. Cancer Res. 25 (20), 6052–6060. PubMed PMID: 31308062.

Tiragolumab Impresses in Multiple Trials, 2020. Cancer Discov. 10 (8), 1086–1087. PubMed PMID: 32576590.

Tosic, V., Thomas, D.L., Kranz, D.M., Liu, J., McFadden, G., Shisler, J.L., et al., 2014. Myxoma virus expressing a fusion protein of interleukin-15 (IL15) and IL15 receptor alpha has enhanced antitumor activity. PLoS One 9 (10), e109801. PubMed PMID: 25329832. Pubmed Central PMCID: 4199602.

Turaj, A.H., Cox, K.L., Penfold, C.A., French, R.R., Mockridge, C.I., Willoughby, J.E., et al., 2018. Augmentation of CD134 (OX40)-dependent NK anti-tumour activity is dependent on antibody cross-linking. Sci. Rep. 8 (1), 2278. PubMed PMID: 29396470. Pubmed Central PMCID: 5797108.

Yu, J., Venstrom, J.M., Liu, X.R., Pring, J., Hasan, R.S., O'Reilly, R.J., et al., 2009. Breaking tolerance to self, circulating natural killer cells expressing inhibitory KIR for non-self HLA exhibit effector function after T cell-depleted allogeneic hematopoietic cell transplantation. Blood 113 (16), 3875–3884. PubMed PMID: 19179302. Pubmed Central PMCID: 2670800.

Yao, X., Matosevic, S., 2021. Chemokine networks modulating natural killer cell trafficking to solid tumors. Cytokine Growth Factor Rev. 59, 36–45. PubMed PMID: 33495094.

Uppendahl, L.D., Felices, M., Bendzick, L., Ryan, C., Kodal, B., Hinderlie, P., et al., 2019. Cytokine-induced memory-like natural killer cells have enhanced function, proliferation, and in vivo expansion against ovarian cancer cells. Gynecol. Oncol. 153 (1), 149–157. PubMed PMID: 30658847. Pubmed Central PMCID: 6430659.

Waldmann, T.A., Dubois, S., Miljkovic, M.D., Conlon, K.C., 2020. IL-15 in the combination immunotherapy of cancer. Front. Immunol. 11, 868. PubMed PMID: 32508818. Pubmed Central PMCID: 7248178.

Wilcox, R.A., Flies, D.B., Zhu, G., Johnson, A.J., Tamada, K., Chapoval, A.I., et al., 2002. Provision of antigen and CD137 signaling breaks immunological ignorance, promoting regression of poorly immunogenic tumors. J. Clin. Invest. 109 (5), 651–659. PubMed PMID: 11877473. Pubmed Central PMCID: 150893.

Vari, F., Arpon, D., Keane, C., Hertzberg, M.S., Talaulikar, D., Jain, S., et al., 2018. Immune evasion via PD-1/PD-L1 on NK cells and monocyte/macrophages is more prominent in Hodgkin lymphoma than DLBCL. Blood 131 (16), 1809–1819. PubMed PMID: 29449276. Pubmed Central PMCID: 5922274.

Zhang, C., Liu, Y., 2020. Targeting NK cell checkpoint receptors or molecules for cancer immunotherapy. Front. Immunol. 11, 1295. PubMed PMID: 32714324. Pubmed Central PMCID: 7344328.

van Montfoort, N., Borst, L., Korrer, M.J., Sluijter, M., Marijt, K.A., Santegoets, S.J., et al., 2018. NKG2A blockade potentiates CD8 T cell immunity induced by cancer vaccines. Cell 175 (7), 1744–1755. e15 PubMed PMID: 30503208. Pubmed Central PMCID: 6354585.

Zhang, X., Sabio, E., Krishna, C., Ma, X., Wang, J., Jiang, H., et al., 2021a. Qa-1(b) modulates resistance to anti-PD-1 immune checkpoint blockade in tumors with defects in antigen processing. Mol. Cancer Res. MCR 19 (6), 1076–1084. PubMed PMID: 33674442. Pubmed Central PMCID: 8178214.

Yeo, J., Ko, M., Lee, D.H., Park, Y., Jin, H.S., 2021. TIGIT/CD226 axis regulates anti-tumor immunity. Pharmaceuticals 14 (3). PubMed PMID: 33670993. Pubmed Central PMCID: 7997242.

Zhang, B., Zhao, W., Li, H., Chen, Y., Tian, H., Li, L., et al., 2016a. Immunoreceptor TIGIT inhibits the cytotoxicity of human cytokine-induced killer cells by interacting with CD155. Cancer Immunol. Immunother. CII 65 (3), 305–314. PubMed PMID: 26842126.

Zhang, C., Wang, H., Li, J., Hou, X., Li, L., Wang, W., et al., 2021b. Involvement of TIGIT in natural killer cell exhaustion and immune escape in patients and mouse model with liver Echinococcus multilocularis infection. Hepatology 74 (6), 3376–3393. PubMed PMID: 34192365.

Vey, N., Karlin, L., Sadot-Lebouvier, S., Broussais, F., Berton-Rigaud, D., Rey, J., et al., 2018. A phase 1 study of lirilumab (antibody against killer immunoglobulin-like receptor antibody KIR2D; IPH2102) in patients with solid tumors and hematologic malignancies. Oncotarget 9 (25), 17675–17688. PubMed PMID: 29707140. Pubmed Central PMCID: 5915148.

Wang, J., Yang, L., Dao, F.T., Wang, Y.Z., Chang, Y., Xu, N., et al., 2022. Prognostic significance of TIM-3 expression pattern at diagnosis in patients with t(8;21) acute myeloid leukemia. Leuk. Lymphoma 63 (1), 152–161. PubMed PMID: 34405769.

Xu, L., Huang, Y., Tan, L., Yu, W., Chen, D., Lu, C., et al., 2015. Increased Tim-3 expression in peripheral NK cells predicts a poorer prognosis and Tim-3 blockade improves NK cell-mediated cytotoxicity in human lung adenocarcinoma. Int. Immunopharmacol. 29 (2), 635–641. PubMed PMID: 26428847.

Yu, L., Liu, X., Wang, X., Yan, F., Wang, P., Jiang, Y., et al., 2021. TIGIT(+) TIM-3(+) NK cells are correlated with NK cell exhaustion and disease progression in patients with hepatitis B virusrelated hepatocellular carcinoma. Oncoimmunology. 10 (1), 1942673. PubMed PMID: 34249476. Pubmed Central PMCID: 8244763.

Wang, Z., Zhu, J., Gu, H., Yuan, Y., Zhang, B., Zhu, D., et al., 2015. The clinical significance of abnormal Tim-3 expression on NK cells from patients with gastric cancer. Immunol. Investig. 44 (6), 578–589. PubMed PMID: 26214042.

Wang, J., Sanmamed, M.F., Datar, I., Su, T.T., Ji, L., Sun, J., et al., 2019. Fibrinogen-like protein 1 is a major immune inhibitory ligand of LAG-3. Cell 176 (1-2), 334–347. e12 PubMed PMID: 30580966. Pubmed Central PMCID: 6365968.

Twomey, J.D., Zhang, B., 2021. Cancer immunotherapy update: FDA-approved checkpoint inhibitors and companion diagnostics. AAPS J. 23 (2), 39. PubMed PMID: 33677681. Pubmed Central PMCID: 7937597.

Xu, C., Marelli, B., Qi, J., Qin, G., Yu, H., Wang, H., et al., 2022. NHS-IL12 and bintrafusp alfa combination therapy enhances antitumor activity in preclinical cancer models. Transl. Oncol. 16, 101322. PubMed PMID: 34954456. Pubmed Central PMCID: 8718653.

Xu, Y., Liu, Q., Zhong, M., Wang, Z., Chen, Z., Zhang, Y., et al., 2019. 2B4 costimulatory domain enhancing cytotoxic ability of anti-CD5 chimeric antigen receptor engineered natural killer cells against T cell malignancies. J. Hematol. Oncol. 12 (1), 49. PubMed PMID: 31097020. Pubmed Central PMCID: PMC6524286. Epub 2019/05/18.

You, F., Wang, Y., Jiang, L., Zhu, X., Chen, D., Yuan, L., et al., 2019. A novel CD7 chimeric antigen receptor-modified NK-92MI cell line targeting T-cell acute lymphoblastic leukemia. Am. J. Cancer Res. 9 (1), 64–78. PubMed PMID: 30755812. Pubmed Central PMCID: PMC6356925. Epub 2019/02/14.

Zhang, Q., Zhang, H., Ding, J., Liu, H., Li, H., Li, H., et al., 2018. Combination therapy with EpCAM-CAR-NK-92 cells and regorafenib against human colorectal cancer models. J Immunol Res 2018, 4263520. PubMed PMID: 30410941. Pubmed Central PMCID: PMC6205314. Epub 2018/11/10.

Zhang, C., Burger, M.C., Jennewein, L., Genssler, S., Schonfeld, K., Zeiner, P., et al., 2016b. ErbB2/HER2-specific NK cells for targeted therapy of glioblastoma. J. Natl. Cancer Inst. 108 (5). PubMed PMID: 26640245. Epub 2015/12/08.

Ueda, T., Kumagai, A., Iriguchi, S., Yasui, Y., Miyasaka, T., Nakagoshi, K., et al., 2020. Non-clinical efficacy, safety and stable clinical cell processing of induced pluripotent stem cell-derived anti-glypican-3 chimeric antigen receptor-expressing natural killer/innate lymphoid cells. Cancer Sci. 111 (5), 1478–1490. PubMed PMID: 32133731. Pubmed Central PMCID: PMC7226201. Epub 2020/03/07.

Zhu, H., Blum, R.H., Bernareggi, D., Ask, E.H., Wu, Z., Hoel, H.J., et al., 2020. Metabolic reprograming via deletion of CISH in human iPSC-derived NK cells promotes in vivo persistence and enhances anti-tumor activity. Cell Stem Cell 27 (2), 224–237. e6 PubMed PMID: 32531207. Pubmed Central PMCID: 7415618.

Xiao, L., Cen, D., Gan, H., Sun, Y., Huang, N., Xiong, H., et al., 2019. Adoptive transfer of NKG2D CAR mRNA-engineered natural killer cells in colorectal cancer patients. Mol. Ther. 27 (6), 1114–1125. PubMed PMID: 30962163. Pubmed Central PMCID: 6554529.

Yun, H.D., Felices, M., Vallera, D.A., Hinderlie, P., Cooley, S., Arock, M., et al., 2018. Trispecific killer engager CD16xIL15xCD33 potently induces NK cell activation and cytotoxicity against neoplastic mast cells. Blood Adv. 2 (13), 1580–1584. PubMed PMID: 29980573. Pubmed Central PMCID: 6039654.

Vallera, D.A., Oh, F., Kodal, B., Hinderlie, P., Geller, M.A., Miller, J.S., et al., 2021. A HER2 tri-specific NK cell engager mediates efficient targeting of human ovarian cancer. Cancer 8, 13(16). PubMed PMID: 34439149. Pubmed Central PMCID: 8394622.

Wu, J., Fu, J., Zhang, M., Liu, D., 2015. AFM13: a first-in-class tetravalent bispecific anti-CD30/CD16A antibody for NK cell-mediated immunotherapy. J. Hematol. Oncol. 8, 96. PubMed PMID: 26231785. Pubmed Central PMCID: 4522136.

Wingert, S., Reusch, U., Knackmuss, S., Kluge, M., Damrat, M., Pahl, J., et al., 2021. Preclinical evaluation of AFM24, a novel CD16A-specific innate immune cell engager targeting EGFR-positive tumors. MAbs 13 (1), 1950264. PubMed PMID: 34325617. Pubmed Central PMCID: 8331026.

Watkins-Yoon, J., Guzman, W., Oliphant, A., Haserlat, S., Leung, A., Chottin, C., et al., 2019. CTX-8573, an innate-cell engager targeting BCMA, is a highly potent multispecific antibody for the treatment of multiple myeloma. Blood 134. PubMed PMID: WOS:000577160408006. English.

Zhang, C., Roder, J., Scherer, A., Bodden, M., Pfeifer Serrahima, J., Bhatti, A., et al., 2021c. Bispecific antibody-mediated redirection of NKG2D-CAR natural killer cells facilitates dual targeting and enhances antitumor activity. J. Immunotherapy Cancer 9 (10). PubMed PMID: 34599028. Pubmed Central PMCID: 8488744.

Zhao, Y., Liu, Z., Li, L., Wu, J., Zhang, H., Zhang, H., et al., 2021. Oncolytic adenovirus: prospects for cancer immunotherapy. Front. Microbiol. 12, 707290. PubMed PMID: 34367111. Pubmed Central PMCID: 8334181.

Uusi-Kerttula, H., Davies, J.A., Thompson, J.M., Wongthida, P., Evgin, L., Shim, K.G., et al., 2018. Ad5NULL-A20: a tropism-modified, alphavbeta6 integrin-selective oncolytic adenovirus for epithelial ovarian cancer therapies. Clin. Cancer Res. 24 (17), 4215–4224. PubMed PMID: 29798908.

Zhang, L., Wang, W., Wang, R., Zhang, N., Shang, H., Bi, Y., et al., 2021d. Reshaping the immune microenvironment by oncolytic herpes simplex virus in murine pancreatic ductal adenocarcinoma. Mol. Ther. 29 (2), 744–761. PubMed PMID: 33130310. Pubmed Central PMCID: 7854309.

Wongthida, P., Diaz, R.M., Galivo, F., Kottke, T., Thompson, J., Pulido, J., et al., 2010. Type III IFN interleukin-28 mediates the antitumor efficacy of oncolytic virus VSV in immune-competent mouse models of cancer. Cancer Res. 70 (11), 4539–4549. PubMed PMID: 20484025. Pubmed Central PMCID: 3896099.

Varudkar, N., Oyer, J.L., Copik, A., Parks, G.D., 2021. Oncolytic parainfluenza virus combines with NK cells to mediate killing of infected and non-infected lung cancer cells within 3D spheroids: role of type I and type III interferon signaling. J. Immunotherapy Cancer 9 (6). PubMed PMID: 34172515. Pubmed Central PMCID: 8237729.

Zhang, J., Tai, L.H., Ilkow, C.S., Alkayyal, A.A., Ananth, A.A., de Souza, C.T., et al., 2014. Maraba MG1 virus enhances natural killer cell function via conventional dendritic cells to reduce postoperative metastatic disease. Mol. Ther. 22 (7), 1320–1332. PubMed PMID: 24695102. Pubmed Central PMCID: 4088996.

Zhang, W., Hu, X., Liang, J., Zhu, Y., Zeng, B., Feng, L., et al., 2020. oHSV2 can target murine colon carcinoma by altering the immune status of the tumor microenvironment and inducing antitumor immunity. Mol. Ther. Oncolytics 16, 158–171. PubMed PMID: 32055679. Pubmed Central PMCID: 7011019.

Zheng, J.N., Pei, D.S., Sun, F.H., Liu, X.Y., Mao, L.J., Zhang, B.F., et al., 2009. Potent antitumor efficacy of interleukin-18 delivered by conditionally replicative adenovirus vector in renal cell carcinoma-bearing nude mice via inhibition of angiogenesis. Cancer Biol. Ther. 8 (7), 599–606. PubMed PMID: 19305163.

Xie, X., Lv, J., Zhu, W., Tian, C., Li, J., Liu, J., et al., 2022. The combination therapy of oncolytic HSV-1 armed with anti-PD-1 antibody and IL-12 enhances anti-tumor efficacy. Transl. Oncol. 15 (1), 101287. PubMed PMID: 34808461. Pubmed Central PMCID: 8607272.

Xu, B., Tian, L., Chen, J., Wang, J., Ma, R., Dong, W., et al., 2021. An oncolytic virus expressing a full-length antibody enhances antitumor innate immune response to glioblastoma. Nat. Commun. 12 (1), 5908. PubMed PMID: 34625564. Pubmed Central PMCID: 8501058.

CHAPTER FOUR

Enabling CAR-T cells for solid tumors: Rage against the suppressive tumor microenvironment

Asier Antoñana-Vildosola[a], Samanta Romina Zanetti[a], and Asis Palazon[a,b,*]

[a]Cancer Immunology and Immunotherapy Lab, CIC bioGUNE, Basque Research and Technology Alliance (BRTA), Bizkaia, Spain
[b]Ikerbasque, Basque Foundation for Science, Bizkaia, Spain
[*]Corresponding author: e-mail address: apalazon@cicbiogune.es

Contents

1. Introduction	124
2. CAR-T cells targeting the tumor vasculature	127
3. CAR-T cells targeting cancer-associated fibroblasts	129
4. CAR-T cells targeting tumor associated macrophages and myeloid suppressor cells	131
5. Concluding remarks	136
Acknowledgments	136
References	136

Abstract

Adoptive T cell therapies based on chimeric antigen receptors (CAR-T) are emerging as genuine therapeutic options for the treatment of hematological malignancies. The observed clinical success has not yet been extended into solid tumor indications as a result of multiple factors including immunosuppressive features of the tumor microenvironment (TME). In this context, an emerging strategy is to design CAR-T cells for the elimination of defined cellular components of the TME, with the objective of re-shaping the tumor immune contexture to control tumor growth. Relevant cell components that are currently under investigation as targets of CAR-T therapies include the tumor vasculature, cancer-associated fibroblasts (CAFs), and immunosuppressive tumor associated macrophages (TAMs) and myeloid derived suppressor cells (MDSCs). In this review, we recapitulate the rapidly expanding field of CAR-T cell therapies that directly target cellular components within the TME with the ultimate objective of promoting immune function, either alone or in combination with other cancer therapies.

1. Introduction

Adoptive cell therapy based on genetic engineering of T cells to express a chimeric antigen receptor (CAR) has emerged on the last decade as a disruptive and efficacious therapeutic strategy. This approach is based on the generation of tumor-targeted T cells by the expression of a synthetic molecule that integrates the recognition specificity of an antibody and the T-cell receptor (TCR) complex signaling. The basic CAR design backbone is comprised of the sequence of an antibody-derived single-chain variable domain (scFv) as an extracellular domain, a transmembrane region and intracellular co-stimulatory and one or more activating signaling domains. Those signaling domains can be introduced in a modular manner, leading to the evolution of different CAR generations based on the incorporation of additional functional decoys (June et al., 2018; June and Sadelain, 2018; Sadelain et al., 2013).

This approach has been validated with autologous primary human T cells, but extensive efforts are currently undergoing to reduce the manufacturing and logistic requirements associated to autologous CAR-T cell therapies. Alternative non-autologous, off-the-shelf cell sources such as natural killer (NK) cells or non-allogeneic T cells would help standardize and deploy this approach on a scalable manner (Rafiq et al., 2020).

The field of CAR-T cell immunotherapy is already consolidated after several FDA approvals. Currently five CAR-T cell products are commercially available for different hematological malignancies, including B-cell acute lymphoblastic leukemia (B-ALL), B cell non-Hodgkin lymphoma (B-NHL), and more recently multiple myeloma (MM) (Abramson et al., 2020; Jacobson et al., 2022; Maude et al., 2018; Munshi et al., 2021; Neelapu et al., 2017; Pasquini et al., 2020; Schuster and Investigators, 2019; Wang et al., 2020).

One of the main challenges of this therapeutic approach is the translation of the success achieved in hematological malignancies into other disease segments, mainly for the treatment of solid tumors. Solid tumors present unique characteristics that pose a challenge, such as heterogeneity in terms of antigen expression. For this reason, multi-targeted CAR strategies are under investigation with the aim of mitigating tumor scape due to lack or downregulation of the tumor antigen (Hou et al., 2021).

The initial attempts of advancing CAR-T cells into solid tumor applications have been focused on the selection of tumor associated antigens (TAA) with the following ideal characteristics: expression on the cell surface, prominent expression on malignant tissue and minimal or lack of expression on healthy organs. An ideal TAA would be expressed on all malignant cells but absent on healthy cells: driving efficacy, preventing relapse, and avoiding off-tumor toxicities.

Currently, several TAAs are under investigation for the treatment of solid tumors (Hou et al., 2021). Among them, TAAs with ongoing clinical trials include: mesothelin (Adusumilli et al., 2021; Morello et al., 2016), GD2 (Mount et al., 2018), HER2 (Vitanza et al., 2021), B7-H3 (Theruvath et al., 2020) and PSMA (Kloss et al., 2018). The challenge in selecting the right TAA for the treatment of solid tumors relies on the fact that CAR-T cell mediated cytotoxicity of functional cells could compromise the physiological function of the organ of origin. In the context of the treatment of solid tumors, severe toxicities have been observed in the clinic, highlighting that the choice of an optimal target antigen represents a critical safety factor (Hou et al., 2021).

Compared to hematological malignancies, solid tumors present additional features that prevent T cell infiltration and efficacy as a consequence of the unique physical barrier associated to the architecture of the tumor mass and the extracellular matrix (ECM). This barrier, together with stromal and immune cell components infiltrating the tumor, and their secreted molecules—such as growth factors, cytokines and chemokines—are termed the tumor microenvironment (TME) (Fig. 1), and influence tumor escape from cell-based immunotherapies (Bagaev et al., 2021; Rodriguez-Garcia et al., 2020).

In this context, an alternative emerging strategy is directly targeting defined elements of the TME, with the objective of re-shaping this environment to a more supportive arena for optimal immune responses, ultimately leading to the control of tumor growth (Bejarano et al., 2021).

Solid tumors have been classified as "hot" or "cold" based on the level of immune infiltration (Galon and Bruni, 2019) and the predicted prognosis and response to immunotherapy (Bruni et al., 2020). In this context, a high mutational and neoantigen burden determines adaptive immune responses (Hellmann et al., 2018; Luksza et al., 2017; Rizvi et al., 2015; Zaretsky et al., 2016), which are required to unleash the full therapeutic potential

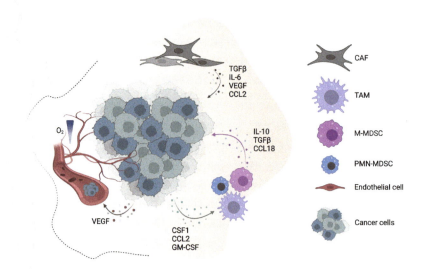

Fig. 1 Integrated view of the tumor microenvironment (TME) components that drive tumor progression and metastasis. Solid tumors present diverse immune and stromal cell components, some of which are resident cells while others are recruited from the periphery. Each of these cell subsets are functionally educated by the tumor and contribute at specific stages of malignant progression. Infiltrating cells include endothelial cells that supply oxygen and nutrients but often proliferate in an aberrant and unfunctional manner; CAFs that directly fuel cancer cell malignancy, promote tumor angiogenesis, and participate in myeloid cell recruitment; and suppressive myeloid cells, that are attracted by soluble factors and acquire immunosuppressive traits, ultimately leading to T cell dysfunction. CAF, cancer-associated fibroblast; CCL2, CC-chemokine ligand 2; CSF1, colony-stimulating factor 1; GM-CSF, granulocyte-monocyte colony stimulating factor; M-MDSC, monocytic myeloid-derived suppressor cell; PMN-MDSC, polymorphonuclear myeloid-derived suppressor cell; TAM, tumor-associated macrophage; TGFβ, transforming growth factor-β.

of checkpoint receptor blockade, a therapy focused on potentiating the function of cytotoxic T cells (Chowell et al., 2018; Engelhard et al., 2018; Tang et al., 2016).

Immune populations present in the TME can be diverse and heterogeneous, differentially influenced by tumor driver mutations and the cancer tissue of origin. Despite this heterogeneity, a recent immune-based classification across different tumor types has identified unique common patterns of infiltrating immune cells, termed archetypes. Archetypes result from the computational clustering of cell compartments classified by expression features and identify the most common immune compositions across cancer patients (Combes et al., 2022). Therefore, such patient stratification might guide therapeutic intervention based on the TME biology.

Apart from adaptive immune populations, myeloid cells infiltrate tumors and have the potential of negatively influencing tumor progression, metastatic success, and response to therapies (Grover et al., 2021; Noy and Pollard, 2014; Veglia et al., 2021). As a result, immunosuppressive myeloid cells are emerging as potential therapeutic targets that could be directly eliminated by CAR-T cells. Other relevant non-immune stromal components of the TME include endothelial cells and fibroblasts.

Angiogenic endothelial cells feed the TME, including immune and malignant cells, providing nutrients and oxygen. In solid tumors, the balance between cell proliferation and vascularization is often disrupted, leading to hypoxia and nutrient deprivation. These can have a detrimental influence on immune responses and promote cancer progression on several tumor types (Carmeliet, 2005). For these reasons, angiogenic endothelial cells are relevant stromal cells susceptible to therapeutic targeting with CAR-T cells.

Cancer-associated fibroblasts (CAFs) are abundant components of the tumor stroma that can also promote tumor growth and therapeutic resistance, making them attractive targets for CAR-T therapy. This strategy is supported by the recent characterization of specific CAF cell surface markers and their classification based on phenotypic and functional diversity (Chen and Song, 2019; Sahai et al., 2020).

Here, we review the rapidly expanding field of CAR-T cell therapies that directly target components of the TME with the ultimate objective of promoting immune function, either alone or in combination with other cancer therapies.

2. CAR-T cells targeting the tumor vasculature

Angiogenic vessels have been considered as therapeutic targets in a wide variety of approaches. Differential features of the TME, such as hypoxia and metabolic changes, often result in aberrant vascularization and tumor progression (Carmeliet and Jain, 2000). Vascular endothelial growth factor (VEGF) is an example of a soluble target that promotes angiogenesis. For this reason, both VEGF and VEGF receptors (VEGFR) have been long considered therapeutic targets that can be blocked by monoclonal antibodies such as bevacizumab (Ferrara et al., 2005). The neutralization of angiogenic signals can promote vascular normalization, a process that restores the biology of endothelial cells from abnormal to more efficient in terms of oxygen, nutrient, and drug delivery (Jain, 2005). A more recent approach consists of cancer vaccines targeting the vasculature after immunization with endothelial antigens (Facciponte et al., 2014).

The endothelial barrier is comprised of endothelial cells supported by pericytes and tumor vascularization can have immunosuppressive effects on tumor infiltrating lymphocytes, a result attributed to compromised nutrient delivery and reduced oxygen availability. To infiltrate tumors, T cells must roll and attach to adhesion molecules expressed on endothelial cells, such as ICAM-1 and VCAM-1 to then extravasate into the tumor (Lanitis et al., 2015). In this process, T cells can be directly inhibited after interaction with inhibitory receptors expressed on endothelial cells, such as PD-L1 (Mazanet and Hughes, 2002).

In this context, the depletion of tumor endothelial cells with CAR-T cells is a promising strategy. The development of CAR-T cell therapies directed against endothelial antigens offers some advantages compared to traditional approaches: genomic stability and accessibility to circulating T cells (Lanitis et al., 2015). Moreover, this approach could be applied to several types of solid tumors. The most advanced endothelial-targeting CAR-T cell candidate is directed against VEGFR, and several studies have demonstrated efficacy in diverse preclinical models (Chinnasamy et al., 2010, 2012; Wang et al., 2013). However, this efficacy is often limited when the CAR-T cells are administered as monotherapy as a result of the binding to circulating VEGF-A to the CAR (Lanitis et al., 2021).

Other molecules considered for CAR-T therapies targeting vascular disruption include: PSMA (Santoro et al., 2015), TEM8 (Petrovic et al., 2019), the EIIIB fibronectin splice variant (Xie et al., 2019), $\alpha v \beta 3$ integrin (Fu et al., 2013), and the NKG2D-ligand Rae1 (Zhang and Sentman, 2013) (Fig. 2). In general, the potential on-target off-tumor effects of these cell therapy candidates remain poorly characterized.

In conclusion, novel cell therapy approaches developed to disrupt the endothelial barrier are a genuine therapeutic modality that offers relevant combinatorial opportunities with other immunotherapies.

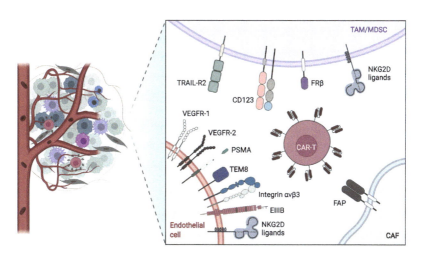

Fig. 2 Microenvironment associated targets undergoing evaluation for CAR-T cell redirection. Depiction of different antigens expressed on stromal and immune components of the TME—including those expressed by TAMs and MDSCs, endothelial cells and CAFs—that are currently being tested for the treatment of solid tumors. CAF, cancer-associated fibroblast; CAR-T, chimeric antigen receptor T cell; EIIIB, alternatively spliced domain of fibronectin-EIIIB; FRβ, folate receptor beta; MDSC, myeloid-derived suppressor cell; PSMA, prostate specific membrane antigen; TAM, tumor-associated macrophage; TEM8, tumor endothelial marker 8; TRAIL-R2, TNF-related apoptosis-inducing ligand receptor 2; VEGFR-1, vascular endothelial growth factor receptor 1; VEGFR-2, vascular endothelial growth factor receptor 2.

3. CAR-T cells targeting cancer-associated fibroblasts

CAFs are one of the most abundant stromal components in the TME and play a key role in tumor initiation and progression (Chen and Song, 2019). Functionally, activated CAFs promote an aberrant ECM deposition and remodeling, hampering the penetrance of T cells to the tumor core and an effective immunosurveillance (Mariathasan et al., 2018; Tauriello et al., 2018; Valkenburg et al., 2018). In the TME, CAFs can also interact with other cell types, stablishing a crosstalk with cancer cells and immune cells that impacts tumor progression and metastatic potential. Based on phenotypic characterization and functional studies, several tumor-promoting functions of CAFs have been described: modulation of epithelial-to-mesenchymal transition (EMT), contribution to metastatic dissemination, and therapeutic resistance (Calon et al., 2012; Vennin et al., 2019; Wang et al., 2016; Yu et al., 2014). In terms of their biology, activated stromal fibroblasts present an altered metabolism that can contribute to chemoresistance through mechanisms like the release of glutathione and cysteine (Wang et al., 2016). CAFs can influence and regulate the infiltration and phenotype of immune cells within the TME, exerting potent immunosuppressive activity as a result of different mechanisms: (i) secretion of chemokines and cytokines (*i.e.*, TGFβ, IL-6, IL-8, CXCL12, CCL2, VEGF), (ii) recruitment of MDSC and monocytes and (iii) induction of a pro-tumoral phenotype in infiltrating macrophages that ultimately limit cytotoxic immune responses (Cho et al., 2018; Feig et al., 2013; Kumar et al., 2017; O'Connell et al., 2011). Notably, the presence of activated CAFs in the TME has been associated with worse clinical prognosis, resistance to therapies, and disease recurrence in multiple cancers (Calon et al., 2015; Finak et al., 2008; Hussain et al., 2020; Ryner et al., 2015; Yamashita et al., 2012).

The central role played by CAFs, both as a physical barrier and as immunosuppressor, sets them as a target to enhance immunotherapy of cancer. Emerging evidence suggests that CAFs are a complex and heterogeneous population of cells based on the origin, phenotypic markers, gene expression profiling, functionality, and plasticity (Bartoschek et al., 2018; Biffi et al., 2019; Elyada et al., 2019; Lambrechts et al., 2018; Pothoff et al., 1992). In this context, under certain circumstances, CAFs can also contribute to cancer progression (Dominguez et al., 2020; Givel et al., 2018; Kieffer et al., 2020).

To date, many candidate markers of CAFs have been identified such as vimentin, α-smooth muscle actin, fibroblast activated protein (FAP), S100A4/fibroblast specific protein 1, periostin, platelet derived growth factor receptor α or β, podoplanin, and integrin β1 among others (Chen et al., 2021). Each marker defines different cell populations that are partially overlapping but are also associated with distinct functional profiles. However, none of these cell surface markers is exclusively expressed by CAFs, adding difficulty to the choice of a safe and restricted antigen for CAR-T cell-mediated depletion of pro-tumoral CAFs in the TME.

The most advanced immune-based treatment strategy developed to deplete CAFs is anti-FAP CAR-T cell therapy (Fig. 2), which has shown mixed results. Several preclinical studies have shown significant anti-tumor efficacy of FAP-targeted CAR-T cells in combination with other antitumor therapies for lung, breast, and pancreatic cancer (Bughda et al., 2021; Kakarla et al., 2013; Lo et al., 2015; Wang et al., 2014; Xin et al., 2021). However, other study using FAP-targeted CAR-T cells reported limited efficacy and lethal cachexia and bone toxicity due to on-target off-tumor toxicities caused by the elimination of FAP$^+$ progenitor stromal cells (Tran et al., 2013). Whether these discrepant results are due to the use of different CAR constructs or other factors merits further investigation. Currently, two clinical trials using anti-FAP CAR-T cells have been conducted. A phase I clinical trial using FAP-CAR-T-cells was initiated in 2012 for the treatment of Malignant Pleural Mesothelioma (NCT01722149), but only 4 patients were recruited (Curioni et al., 2019; Petrausch et al., 2012; Pircher et al., 2015). Another recent clinical trial aims to use a fourth-generation CAR-T targeting Nectin4/FAP for advanced malignant solid tumors (NCT03932565).

Despite numerous advances in the understanding of the biology of CAFs, the current lack of specific CAF cell markers along with the heterogeneity of CAFs limits the translation of CAR-T cell therapy to eliminate CAFs in the TME.

4. CAR-T cells targeting tumor associated macrophages and myeloid suppressor cells

A common feature of most solid tumors, including those with low T cell infiltration, is the presence of infiltrating myeloid cells. This compartment of the TME encompasses heterogeneous cell subsets with a dominant immunosuppressive role (Engblom et al., 2016). Suppressive myeloid populations promote tumor progression by blunting antitumor immunity through direct and indirect mechanisms in breast (Cassetta et al., 2019; DeNardo et al., 2009; Medrek et al., 2012; Williams et al., 2016), pancreatic (Bear et al., 2020), hepatocellular (Hoechst et al., 2008; Wan et al., 2015), and prostate cancer (Lopez-Bujanda and Drake, 2017), among others. Importantly, the presence of tumor associated myeloid cell populations has been linked with metastasis and poor clinical outcome in numerous disease segments (Hutchinson et al., 2020; Qian et al., 2011; Qian and Pollard, 2010; Steidl et al., 2010; Szekely et al., 2018; Zhu et al., 2019).

Historically, classification of myeloid cells in the context of cancer has been complex and based on the expression of a discrete number of surface markers that can be identified by traditional flow cytometry (Gabrilovich et al., 2012). More recently, technological advances have allowed for a more comprehensive classification based on single-cell RNA sequencing and mass-cytometry expression profiles across several tumor types (Cheng et al., 2021; Gubin et al., 2018), further evidencing the genuine heterogeneity of the myeloid compartment within the TME.

Phenotypic plasticity is one of the hallmarks of myeloid cells (DeNardo and Ruffell, 2019). The tumor context shapes myeloid polarization through a diverse variety of signals mediated by cytokines, chemokines, metabolites and soluble factors released from surrounding cells (Biswas and Mantovani, 2010; Mantovani et al., 2017). Oxygen availability is often limited in solid tumors, contributing to the phenotypic heterogeneity of myeloid cells that are exposed to different levels of hypoxia depending on their spatial location within the tumor mass (Casazza et al., 2014). Hypoxia can contribute to the recruitment of myeloid cells by promoting the release of recruitment factors such as CCL2 and CSF1. Moreover, HIF activity can influence polarization, secretion of cytokines and tumor infiltration (Corzo et al., 2010; DeNardo and Ruffell, 2019; Henze and Mazzone, 2016).

Despite intrinsic plasticity and tumor-induced phenotypic diversity, both at different regions of a tumor and between different tumors, infiltrating myeloid cells are mostly comprised of immunosuppressive tumor associated macrophages (TAMs), myeloid-derived suppressor cells (MDSCs) and dendritic cells (DCs) (Binnewies et al., 2018).

Macrophages are defined as differentiated cells that belong to the mononuclear phagocytic lineage. In fact, TAMs are one of the most abundant immune infiltrates in many tumors and have a dominant role in tumor initiation and promotion of metastasis (Noy and Pollard, 2014; Qian and Pollard, 2010).

Within the TME, TAMs can arise from 2 different origins: monocyte-derived macrophages (MDMs), which are newly recruited and maintained by the adult hematopoietic system; and tissue-resident macrophages (TRMs), consisting of a self-renewing population that mainly arises from embryonic progenitors already present in the tissue where the tumor originates (Bleriot et al., 2020; Epelman et al., 2014).

As our understanding of the diversity and biology of TAMs grows, it is becoming clear that MDMs and TRMs contribute to tumor progression at different levels. At the early stages of malignancy, TRMs facilitate the establishment and progression of the primary tumor and can promote metastatic dissemination (Casanova-Acebes et al., 2021; Etzerodt et al., 2020; Loyher et al., 2018; Yona et al., 2013; Zhu et al., 2017). Also, the tumor mass can attract circulating monocytes mainly through the CCL2–CCR2 and CSF1–CSF1R signaling pathways. As a result, recruited monocytes differentiate into MDMs that ultimately facilitate tumor metastasis (Consonni et al., 2021; DeNardo et al., 2009; Nagarsheth et al., 2017; Ozga et al., 2021; Qian et al., 2011).

Other predominant component of the TME are MDSCs, which are neutrophils and monocytes present in the TME in an activated state that inhibit cytotoxic immune responses (Veglia et al., 2021). Currently, two major groups of MDSCs have been characterized based on their cell of origin: monocytic MDSCs (M-MDSCs) and polymorphonuclear MDSCs (PMN-MDSCs).

Myeloid cells promote tumor growth *via* both non-immune and immune-mediated mechanisms. Non-immune mechanisms that promote cancer cell malignancy are multi-factorial and diverse. One example is the contribution of macrophages to EMT, which in turn accentuates malignant and stem cell traits in cancer cells (Bonde et al., 2012; Sharma et al., 2021; Su et al., 2014; Wan et al., 2014; Williams et al., 2016). Additionally, TAMs and MDSCs promote the metastatic process (Acharyya et al., 2012; Chen et al., 2011; DeNardo et al., 2009; Kitamura et al., 2015; Pollard, 2004; Steele et al., 2016), resistance to chemotherapy (Acharyya et al., 2012; DeNardo et al., 2011; Yin et al., 2017), and angiogenesis (De Palma et al., 2017; Riabov et al., 2014).

The most studied mechanism by which myeloid cells directly promote immune evasion is the one mediated by inhibitory ligands of checkpoint receptors. Some of the relevant immune checkpoint ligands mediating the myeloid–T cell crosstalk within the TME include PD-L1 (Kuang et al., 2009; Lin et al., 2018), B7-H4 (Kryczek et al., 2006; Li et al., 2018), VISTA (Li et al., 2022), TIM4 (Chow et al., 2021), and the family of Sialic-acid binding domains (SIGLECs) (Barkal et al., 2019; Beatson et al., 2016; Rodriguez et al., 2021; Wang et al., 2019), among many others.

Another mechanism by which myeloid cells directly suppress cytotoxic T lymphocyte functions is by the secretion of immunoregulatory cytokines such as IL-10 (Ruffell et al., 2014; Smith et al., 2018) and TGFβ (Mariathasan et al., 2018; Movahedi et al., 2010; Tauriello et al., 2018; Torroella-Kouri et al., 2009). In the case of IL-10, additional indirect mechanisms of T cell suppression have been described (Ruffell et al., 2014). TAMs can also secret chemokines (*i.e.*, CCL22 and CCL20) that hinder T cell responses indirectly through the recruitment of regulatory T cells to the tumor bed (Curiel et al., 2004; Wu et al., 2019). For these reasons, cytokines and chemokines offer alternative therapeutic strategies (Propper and Balkwill, 2022).

Mounting evidence indicates that anti-inflammatory and suppressor myeloid cells play a critical role in regulating different steps of tumor progression and metastasis. In this context, approaches that aim to deplete or re-educate these cell populations are of relevant interest (Cassetta and Pollard, 2018). These include targeting TAMs with anti-CSF1R effector antibodies, a drug that has been evaluated for multiple cancer indications including pancreatic ductal adenocarcinoma (PDAC), colorectal cancer (CRC) and triple-negative breast cancer (TNBC)—either alone or combined with other therapeutic agents (Cassetta and Kitamura, 2018; Cassetta and Pollard, 2018; Mehta et al., 2021; Neubert et al., 2018; Ries et al., 2014; Salvagno et al., 2019; Stromnes et al., 2019).

Despite encouraging experimental and clinical results, a potential drawback of this approach lays on the fact that some of the myeloid cells recruited to the TME have relevant anti-tumor functions through antigen presentation, co-stimulation, and epitope spreading (Asano et al., 2011; Forssell et al., 2007; Kuang et al., 2010; Ong et al., 2012; Pituch et al., 2018; Stromnes et al., 2019). As a result, pan-myeloid depletion in the tumor microenvironment *via* anti-CSF1R or similar strategies may be less beneficial than reprogramming the myeloid compartment either through inducing a phenotype switch or by specific depletion of defined suppressive myeloid subsets.

Because most markers expressed by immunosuppressive TAMs and MDSCs are shared with anti-tumor or proinflammatory myeloid subpopulations, the choice of selective markers that allow specific depletion of suppressive subsets remains one of the main challenges in this field.

Several adoptive transfer therapies based on CAR-T cells targeting myeloid cells have been tested at preclinical stage (Fig. 2). One of the first targets proposed for targeting the immunosuppressive TME of Hodgkin Lymphoma (HL) was CD123 (Ruella et al., 2017). CD123, a target under clinical development for acute myeloid leukemia (AML), is expressed by both HL cancer cells and immunosuppressive TAMs. However, despite experimental data supporting the success of the strategy, depleting $CD123^+$ cells might result in life-threatening toxicities, given the expression of CD123 on healthy immature and myeloid hematopoietic stem/progenitor cells (HSPCs). This broad pattern of CD123 expression raises concerns related to potential on-target off-tumor myeloablative toxicities after anti-CD123 CAR-T cell immunotherapy (Baroni et al., 2020; Gill et al., 2014).

To minimize risks associated with pan-myeloid targeting strategies, an emerging effort involves the identification of markers that are differentially and specifically expressed by immunosuppressive myeloid cells.

An example of such a marker is the family of TNF-related apoptosis-induced ligand receptors (TRAIL-Rs). MDSCs have prominent expression of TRAIL-Rs (Condamine et al., 2014), and administration of agonistic anti-TRAIL-R2 antibodies are largely well tolerated in the clinic. Indeed, treatment does not deplete normal mature myeloid or lymphoid cells and the reduction in MDSCs numbers correlates with a longer progression-free survival (Dominguez et al., 2017). Based on this data, a novel CAR-T cell immunotherapeutic strategy to target both cancer cells and MDSCs has been proposed. The approach is based on a conventional anti-MUC1 tumor-targeted CAR equipped with a costimulatory receptor which comprises the scFv of an anti-TRAIL-R2 agonistic antibody and a 4-1BB intracellular signaling domain (Nalawade et al., 2021). The aim of this strategy is to genetically modify T cells to target MUC1-expressing breast cancer cells, induce apoptosis of MDSCs, and provide a second costimulatory signal for the optimal activation of the T cells. This example illustrates the variety of CAR strategies that are being tested at the preclinical level, including those that directly modulate the biology of myeloid cells.

In order to minimize potential toxicities associated with off-tumor expression of the antigen of choice, alternative CAR-based cell platforms can be engineered to limit their action within the TME. For example, NK cells can be armored with a NKG2D-based CAR to deplete MDSCs and administered in combination with anti-GD2 CAR-T cells redirected against cancer cells (Parihar et al., 2019). This experimental approach showed preclinical efficacy in neuroblastoma models while limiting the toxicity previously seen when a NKG2D CAR receptor was expressed on T cells (VanSeggelen et al., 2015).

The folate receptor beta (FRβ) is other antigen expressed by immunosuppressive TAMs in various cancer types (O'Shannessy et al., 2015; Puig-Kroger et al., 2009; Shen et al., 2015). A recent approach directly targeted TAMs with anti-FRβ CAR-T cells for the treatment of ovarian cancer. This strategy showed promising results when administered as pre-conditioning regimen before the treatment with CAR-T cells targeting TAAs, such as mesothelin (Rodriguez-Garcia et al., 2021).

Altogether, targeting myeloid cells with adoptive cell therapies is a promising emerging field given their pivotal role in the progression and resistance to immune checkpoint blockade and chemotherapy. Despite advances in the characterization of infiltrating myeloid cells, a deeper understanding of myeloid subtypes and identification of specific markers is required for the design of the next generation therapies.

5. Concluding remarks

Given their role in determining tumor aggressiveness and metastatic potential, the varied stromal and immune cellular components within the TME are relevant factors that influence the failure, response, or resistance to cancer therapies. Tumors are heterogeneous in terms of cell composition, and each TME population differentially contributes to the clinical outcomes of classical therapies, immune checkpoint inhibitors and adoptive T cell therapies. An emerging approach consists of directly targeting these cellular components of the TME by selecting appropriate antigens for CAR-T-mediated cell depletion. CAR-T cells targeting tumor infiltrating endothelial cells, fibroblasts or immunosuppressive myeloid cells may reshape the tumor landscape and increase the therapeutic index of other immunotherapies for solid tumors when administered in combination.

Acknowledgments

A.P. receives funding from: European Research Council (ERC) (ERC-2018-StG 804236-NEXTGEN-IO); AECClab (LABAE211744PALA), La Caixa Health Research (LCF/PR/HR21/52410009); Fundación Fero and Ikerbasque. Proyecto PID2019-107956RA-I00 financiado por MCIN/AEI/10.13039/501100011033/. Ayuda RYC2018-024183-I financiada por MCIN/AEI/10.13039/501100011033 y por El FSE invierte en tu futuro. A.A.-V. receives funding from "la Caixa" Foundation (ID 100010434, LCF/BQ/DR20/11790022).

References

Abramson, J.S., Palomba, M.L., Gordon, L.I., Lunning, M.A., Wang, M., Arnason, J., Mehta, A., Purev, E., Maloney, D.G., Andreadis, C., et al., 2020. Lisocabtagene maraleucel for patients with relapsed or refractory large B-cell lymphomas (TRANSCEND NHL 001): a multicentre seamless design study. Lancet 396, 839–852.

Acharyya, S., Oskarsson, T., Vanharanta, S., Malladi, S., Kim, J., Morris, P.G., Manova-Todorova, K., Leversha, M., Hogg, N., Seshan, V.E., et al., 2012. A CXCL1 paracrine network links cancer chemoresistance and metastasis. Cell 150, 165–178.

Adusumilli, P.S., Zauderer, M.G., Riviere, I., Solomon, S.B., Rusch, V.W., O'Cearbhaill, R.E., Zhu, A., Cheema, W., Chintala, N.K., Halton, E., et al., 2021. A phase I trial of regional mesothelin-targeted CAR T-cell therapy in patients with malignant pleural disease, in combination with the anti-PD-1 agent pembrolizumab. Cancer Discov. 11, 2748–2763.

Asano, K., Nabeyama, A., Miyake, Y., Qiu, C.H., Kurita, A., Tomura, M., Kanagawa, O., Fujii, S., Tanaka, M., 2011. CD169-positive macrophages dominate antitumor immunity by crosspresenting dead cell-associated antigens. Immunity 34, 85–95.

Bagaev, A., Kotlov, N., Nomie, K., Svekolkin, V., Gafurov, A., Isaeva, O., Osokin, N., Kozlov, I., Frenkel, F., Gancharova, O., et al., 2021. Conserved pan-cancer microenvironment subtypes predict response to immunotherapy. Cancer Cell 39, 845–865. e847.

Barkal, A.A., Brewer, R.E., Markovic, M., Kowarsky, M., Barkal, S.A., Zaro, B.W., Krishnan, V., Hatakeyama, J., Dorigo, O., Barkal, L.J., Weissman, I.L., 2019. CD24 signalling through macrophage Siglec-10 is a target for cancer immunotherapy. Nature 572, 392–396.

Baroni, M.L., Sanchez Martinez, D., Gutierrez Aguera, F., Roca Ho, H., Castella, M., Zanetti, S.R., Velasco Hernandez, T., Diaz de la Guardia, R., Castano, J., Anguita, E., et al., 2020. 41BB-based and CD28-based CD123-redirected T-cells ablate human normal hematopoiesis in vivo. J. Immunother. Cancer 8.

Bartoschek, M., Oskolkov, N., Bocci, M., Lovrot, J., Larsson, C., Sommarin, M., Madsen, C.D., Lindgren, D., Pekar, G., Karlsson, G., et al., 2018. Spatially and functionally distinct subclasses of breast cancer-associated fibroblasts revealed by single cell RNA sequencing. Nat. Commun. 9, 5150.

Bear, A.S., Vonderheide, R.H., O'Hara, M.H., 2020. Challenges and opportunities for pancreatic cancer immunotherapy. Cancer Cell 38, 788–802.

Beatson, R., Tajadura-Ortega, V., Achkova, D., Picco, G., Tsourouktsoglou, T.D., Klausing, S., Hillier, M., Maher, J., Noll, T., Crocker, P.R., et al., 2016. The mucin MUC1 modulates the tumor immunological microenvironment through engagement of the lectin Siglec-9. Nat. Immunol. 17, 1273–1281.

Bejarano, L., Jordao, M.J.C., Joyce, J.A., 2021. Therapeutic targeting of the tumor microenvironment. Cancer Discov. 11, 933–959.

Biffi, G., Oni, T.E., Spielman, B., Hao, Y., Elyada, E., Park, Y., Preall, J., Tuveson, D.A., 2019. IL1-induced JAK/STAT signaling is antagonized by TGFbeta to shape CAF heterogeneity in pancreatic ductal adenocarcinoma. Cancer Discov. 9, 282–301.

Binnewies, M., Roberts, E.W., Kersten, K., Chan, V., Fearon, D.F., Merad, M., Coussens, L.M., Gabrilovich, D.I., Ostrand-Rosenberg, S., Hedrick, C.C., et al., 2018. Understanding the tumor immune microenvironment (TIME) for effective therapy. Nat. Med. 24, 541–550.

Biswas, S.K., Mantovani, A., 2010. Macrophage plasticity and interaction with lymphocyte subsets: cancer as a paradigm. Nat. Immunol. 11, 889–896.

Bleriot, C., Chakarov, S., Ginhoux, F., 2020. Determinants of resident tissue macrophage identity and function. Immunity 52, 957–970.

Bonde, A.K., Tischler, V., Kumar, S., Soltermann, A., Schwendener, R.A., 2012. Intratumoral macrophages contribute to epithelial-mesenchymal transition in solid tumors. BMC Cancer 12, 35.

Bruni, D., Angell, H.K., Galon, J., 2020. The immune contexture and immunoscore in cancer prognosis and therapeutic efficacy. Nat. Rev. Cancer 20, 662–680.

Bughda, R., Dimou, P., D'Souza, R.R., Klampatsa, A., 2021. Fibroblast activation protein (FAP)-targeted CAR-T cells: launching an attack on tumor stroma. ImmunoTargets Ther. 10, 313–323.

Calon, A., Espinet, E., Palomo-Ponce, S., Tauriello, D.V., Iglesias, M., Cespedes, M.V., Sevillano, M., Nadal, C., Jung, P., Zhang, X.H., et al., 2012. Dependency of colorectal cancer on a TGF-beta-driven program in stromal cells for metastasis initiation. Cancer Cell 22, 571–584.

Calon, A., Lonardo, E., Berenguer-Llergo, A., Espinet, E., Hernando-Momblona, X., Iglesias, M., Sevillano, M., Palomo-Ponce, S., Tauriello, D.V., Byrom, D., et al., 2015. Stromal gene expression defines poor-prognosis subtypes in colorectal cancer. Nat. Genet. 47, 320–329.

Carmeliet, P., 2005. Angiogenesis in life, disease and medicine. Nature 438, 932–936.

Carmeliet, P., Jain, R.K., 2000. Angiogenesis in cancer and other diseases. Nature 407, 249–257.

Casanova-Acebes, M., Dalla, E., Leader, A.M., LeBerichel, J., Nikolic, J., Morales, B.M., Brown, M., Chang, C., Troncoso, L., Chen, S.T., et al., 2021. Tissue-resident macrophages provide a pro-tumorigenic niche to early NSCLC cells. Nature 595, 578–584.

Casazza, A., Di Conza, G., Wenes, M., Finisguerra, V., Deschoemaeker, S., Mazzone, M., 2014. Tumor stroma: a complexity dictated by the hypoxic tumor microenvironment. Oncogene 33, 1743–1754.

Cassetta, L., Kitamura, T., 2018. Targeting tumor-associated macrophages as a potential strategy to enhance the response to immune checkpoint inhibitors. Front. Cell Dev. Biol. 6, 38.

Cassetta, L., Pollard, J.W., 2018. Targeting macrophages: therapeutic approaches in cancer. Nat. Rev. Drug Discov. 17, 887–904.

Cassetta, L., Fragkogianni, S., Sims, A.H., Swierczak, A., Forrester, L.M., Zhang, H., Soong, D.Y.H., Cotechini, T., Anur, P., Lin, E.Y., et al., 2019. Human tumor-associated macrophage and monocyte transcriptional landscapes reveal cancer-specific reprogramming, biomarkers, and therapeutic targets. Cancer Cell 35, 588–602. e510.

Chen, X., Song, E., 2019. Turning foes to friends: targeting cancer-associated fibroblasts. Nat. Rev. Drug Discov. 18, 99–115.

Chen, J., Yao, Y., Gong, C., Yu, F., Su, S., Chen, J., Liu, B., Deng, H., Wang, F., Lin, L., et al., 2011. CCL18 from tumor-associated macrophages promotes breast cancer metastasis via PITPNM3. Cancer Cell 19, 541–555.

Chen, P.Y., Wei, W.F., Wu, H.Z., Fan, L.S., Wang, W., 2021. Cancer-associated fibroblast heterogeneity: a factor that cannot be ignored in immune microenvironment remodeling. Front. Immunol. 12, 671595.

Cheng, S., Li, Z., Gao, R., Xing, B., Gao, Y., Yang, Y., Qin, S., Zhang, L., Ouyang, H., Du, P., et al., 2021. A pan-cancer single-cell transcriptional atlas of tumor infiltrating myeloid cells. Cell 184, 792–809. e723.

Chinnasamy, D., Yu, Z., Theoret, M.R., Zhao, Y., Shrimali, R.K., Morgan, R.A., Feldman, S.A., Restifo, N.P., Rosenberg, S.A., 2010. Gene therapy using genetically modified lymphocytes targeting VEGFR-2 inhibits the growth of vascularized syngenic tumors in mice. J. Clin. Invest. 120, 3953–3968.

Chinnasamy, D., Yu, Z., Kerkar, S.P., Zhang, L., Morgan, R.A., Restifo, N.P., Rosenberg, S.A., 2012. Local delivery of interleukin-12 using T cells targeting VEGF receptor-2 eradicates multiple vascularized tumors in mice. Clin. Cancer Res. 18, 1672–1683.

Cho, H., Seo, Y., Loke, K.M., Kim, S.W., Oh, S.M., Kim, J.H., Soh, J., Kim, H.S., Lee, H., Kim, J., et al., 2018. Cancer-stimulated CAFs enhance monocyte differentiation and protumoral TAM activation via IL6 and GM-CSF secretion. Clin. Cancer Res. 24, 5407–5421.

Chow, A., Schad, S., Green, M.D., Hellmann, M.D., Allaj, V., Ceglia, N., Zago, G., Shah, N.S., Sharma, S.K., Mattar, M., et al., 2021. Tim-4(+) cavity-resident macrophages impair anti-tumor CD8(+) T cell immunity. Cancer Cell 39, 973–988. e979.

Chowell, D., Morris, L.G.T., Grigg, C.M., Weber, J.K., Samstein, R.M., Makarov, V., Kuo, F., Kendall, S.M., Requena, D., Riaz, N., et al., 2018. Patient HLA class I genotype influences cancer response to checkpoint blockade immunotherapy. Science 359, 582–587.

Combes, A.J., Samad, B., Tsui, J., Chew, N.W., Yan, P., Reeder, G.C., Kushnoor, D., Shen, A., Davidson, B., Barczak, A.J., et al., 2022. Discovering dominant tumor immune archetypes in a pan-cancer census. Cell 185, 184–203. e119.

Condamine, T., Kumar, V., Ramachandran, I.R., Youn, J.I., Celis, E., Finnberg, N., El-Deiry, W.S., Winograd, R., Vonderheide, R.H., English, N.R., et al., 2014. ER stress regulates myeloid-derived suppressor cell fate through TRAIL-R-mediated apoptosis. J. Clin. Invest. 124, 2626–2639.

Consonni, F.M., Bleve, A., Totaro, M.G., Storto, M., Kunderfranco, P., Termanini, A., Pasqualini, F., Ali, C., Pandolfo, C., Sgambelluri, F., et al., 2021. Heme catabolism by tumor-associated macrophages controls metastasis formation. Nat. Immunol. 22, 595–606.

Corzo, C.A., Condamine, T., Lu, L., Cotter, M.J., Youn, J.I., Cheng, P., Cho, H.I., Celis, E., Quiceno, D.G., Padhya, T., et al., 2010. HIF-1alpha regulates function and differentiation of myeloid-derived suppressor cells in the tumor microenvironment. J. Exp. Med. 207, 2439–2453.

Curiel, T.J., Coukos, G., Zou, L., Alvarez, X., Cheng, P., Mottram, P., Evdemon-Hogan, M., Conejo-Garcia, J.R., Zhang, L., Burow, M., et al., 2004. Specific recruitment of regulatory T cells in ovarian carcinoma fosters immune privilege and predicts reduced survival. Nat. Med. 10, 942–949.

Curioni, A., Britschgi, C., Hiltbrunner, S., Bankel, L., Gulati, P., Weder, W., Opitz, I., Lauk, O., Caviezel, C., Knuth, A., et al., 2019. 1226P—a phase I clinical trial of malignant pleural mesothelioma treated with locally delivered autologous anti-FAP-targeted CAR T-cells. Ann. Oncol. 30, v501.

De Palma, M., Biziato, D., Petrova, T.V., 2017. Microenvironmental regulation of tumour angiogenesis. Nat. Rev. Cancer 17, 457–474.

DeNardo, D.G., Ruffell, B., 2019. Macrophages as regulators of tumour immunity and immunotherapy. Nat. Rev. Immunol. 19, 369–382.

DeNardo, D.G., Barreto, J.B., Andreu, P., Vasquez, L., Tawfik, D., Kolhatkar, N., Coussens, L.M., 2009. CD4(+) T cells regulate pulmonary metastasis of mammary carcinomas by enhancing protumor properties of macrophages. Cancer Cell 16, 91–102.

DeNardo, D.G., Brennan, D.J., Rexhepaj, E., Ruffell, B., Shiao, S.L., Madden, S.F., Gallagher, W.M., Wadhwani, N., Keil, S.D., Junaid, S.A., et al., 2011. Leukocyte complexity predicts breast cancer survival and functionally regulates response to chemotherapy. Cancer Discov. 1, 54–67.

Dominguez, G.A., Condamine, T., Mony, S., Hashimoto, A., Wang, F., Liu, Q., Forero, A., Bendell, J., Witt, R., Hockstein, N., et al., 2017. Selective targeting of myeloid-derived suppressor cells in cancer patients using DS-8273a, an agonistic TRAIL-R2 antibody. Clin. Cancer Res. 23, 2942–2950.

Dominguez, C.X., Muller, S., Keerthivasan, S., Koeppen, H., Hung, J., Gierke, S., Breart, B., Foreman, O., Bainbridge, T.W., Castiglioni, A., et al., 2020. Single-cell RNA sequencing reveals stromal evolution into LRRC15(+) myofibroblasts as a determinant of patient response to cancer immunotherapy. Cancer Discov. 10, 232–253.

Elyada, E., Bolisetty, M., Laise, P., Flynn, W.F., Courtois, E.T., Burkhart, R.A., Teinor, J.A., Belleau, P., Biffi, G., Lucito, M.S., et al., 2019. Cross-species single-cell analysis of pancreatic ductal adenocarcinoma reveals antigen-presenting cancer-associated fibroblasts. Cancer Discov. 9, 1102–1123.

Engblom, C., Pfirschke, C., Pittet, M.J., 2016. The role of myeloid cells in cancer therapies. Nat. Rev. Cancer 16, 447–462.

Engelhard, V.H., Rodriguez, A.B., Mauldin, I.S., Woods, A.N., Peske, J.D., Slingluff Jr., C.L., 2018. Immune cell infiltration and tertiary lymphoid structures as determinants of antitumor immunity. J. Immunol. 200, 432–442.

Epelman, S., Lavine, K.J., Beaudin, A.E., Sojka, D.K., Carrero, J.A., Calderon, B., Brija, T., Gautier, E.L., Ivanov, S., Satpathy, A.T., et al., 2014. Embryonic and adult-derived resident cardiac macrophages are maintained through distinct mechanisms at steady state and during inflammation. Immunity 40, 91–104.

Etzerodt, A., Moulin, M., Doktor, T.K., Delfini, M., Mossadegh-Keller, N., Bajenoff, M., Sieweke, M.H., Moestrup, S.K., Auphan-Anezin, N., Lawrence, T., 2020. Tissue-resident macrophages in omentum promote metastatic spread of ovarian cancer. J. Exp. Med. 217.

Facciponte, J.G., Ugel, S., De Sanctis, F., Li, C., Wang, L., Nair, G., Sehgal, S., Raj, A., Matthaiou, E., Coukos, G., Facciabene, A., 2014. Tumor endothelial marker 1-specific DNA vaccination targets tumor vasculature. J. Clin. Invest. 124, 1497–1511.

Feig, C., Jones, J.O., Kraman, M., Wells, R.J., Deonarine, A., Chan, D.S., Connell, C.M., Roberts, E.W., Zhao, Q., Caballero, O.L., et al., 2013. Targeting CXCL12 from FAP-expressing carcinoma-associated fibroblasts synergizes with anti-PD-L1 immunotherapy in pancreatic cancer. Proc. Natl. Acad. Sci. U. S. A. 110, 20212–20217.

Ferrara, N., Hillan, K.J., Novotny, W., 2005. Bevacizumab (Avastin), a humanized anti-VEGF monoclonal antibody for cancer therapy. Biochem. Biophys. Res. Commun. 333, 328–335.

Finak, G., Bertos, N., Pepin, F., Sadekova, S., Souleimanova, M., Zhao, H., Chen, H., Omeroglu, G., Meterissian, S., Omeroglu, A., et al., 2008. Stromal gene expression predicts clinical outcome in breast cancer. Nat. Med. 14, 518–527.

Forssell, J., Oberg, A., Henriksson, M.L., Stenling, R., Jung, A., Palmqvist, R., 2007. High macrophage infiltration along the tumor front correlates with improved survival in colon cancer. Clin. Cancer Res. 13, 1472–1479.

Fu, X., Rivera, A., Tao, L., Zhang, X., 2013. Genetically modified T cells targeting neovasculature efficiently destroy tumor blood vessels, shrink established solid tumors and increase nanoparticle delivery. Int. J. Cancer 133, 2483–2492.

Gabrilovich, D.I., Ostrand-Rosenberg, S., Bronte, V., 2012. Coordinated regulation of myeloid cells by tumours. Nat. Rev. Immunol. 12, 253–268.

Galon, J., Bruni, D., 2019. Approaches to treat immune hot, altered and cold tumours with combination immunotherapies. Nat. Rev. Drug Discov. 18, 197–218.

Gill, S., Tasian, S.K., Ruella, M., Shestova, O., Li, Y., Porter, D.L., Carroll, M., Danet-Desnoyers, G., Scholler, J., Grupp, S.A., et al., 2014. Preclinical targeting of human acute myeloid leukemia and myeloablation using chimeric antigen receptor-modified T cells. Blood 123, 2343–2354.

Givel, A.M., Kieffer, Y., Scholer-Dahirel, A., Sirven, P., Cardon, M., Pelon, F., Magagna, I., Gentric, G., Costa, A., Bonneau, C., et al., 2018. miR200-regulated CXCL12beta promotes fibroblast heterogeneity and immunosuppression in ovarian cancers. Nat. Commun. 9, 1056.

Grover, A., Sanseviero, E., Timosenko, E., Gabrilovich, D.I., 2021. Myeloid-derived suppressor cells: a propitious road to clinic. Cancer Discov. 11, 2693–2706.

Gubin, M.M., Esaulova, E., Ward, J.P., Malkova, O.N., Runci, D., Wong, P., Noguchi, T., Arthur, C.D., Meng, W., Alspach, E., et al., 2018. High-dimensional analysis delineates myeloid and lymphoid compartment remodeling during successful immune-checkpoint cancer therapy. Cell 175, 1014–1030. e1019.

Hellmann, M.D., Callahan, M.K., Awad, M.M., Calvo, E., Ascierto, P.A., Atmaca, A., Rizvi, N.A., Hirsch, F.R., Selvaggi, G., Szustakowski, J.D., et al., 2018. Tumor mutational burden and efficacy of nivolumab monotherapy and in combination with ipilimumab in small-cell lung cancer. Cancer Cell 33, 853–861. e854.

Henze, A.T., Mazzone, M., 2016. The impact of hypoxia on tumor-associated macrophages. J. Clin. Invest. 126, 3672–3679.

Hoechst, B., Ormandy, L.A., Ballmaier, M., Lehner, F., Kruger, C., Manns, M.P., Greten, T.F., Korangy, F., 2008. A new population of myeloid-derived suppressor cells in hepatocellular carcinoma patients induces CD4(+)CD25(+)Foxp3(+) T cells. Gastroenterology 135, 234–243.

Hou, A.J., Chen, L.C., Chen, Y.Y., 2021. Navigating CAR-T cells through the solid-tumour microenvironment. Nat. Rev. Drug Discov. 20, 531–550.

Hussain, A., Voisin, V., Poon, S., Karamboulas, C., Bui, N.H.B., Meens, J., Dmytryshyn, J., Ho, V.W., Tang, K.H., Paterson, J., et al., 2020. Distinct fibroblast functional states drive clinical outcomes in ovarian cancer and are regulated by TCF21. J. Exp. Med. 217.

Hutchinson, K.E., Yost, S.E., Chang, C.W., Johnson, R.M., Carr, A.R., McAdam, P.R., Halligan, D.L., Chang, C.C., Schmolze, D., Liang, J., Yuan, Y., 2020. Comprehensive profiling of poor-risk paired primary and recurrent triple-negative breast cancers reveals immune phenotype shifts. Clin. Cancer Res. 26, 657–668.

Jacobson, C.A., Chavez, J.C., Sehgal, A.R., William, B.M., Munoz, J., Salles, G., Munshi, P.N., Casulo, C., Maloney, D.G., de Vos, S., et al., 2022. Axicabtagene ciloleucel in relapsed or refractory indolent non-Hodgkin lymphoma (ZUMA-5): a single-arm, multicentre, phase 2 trial. Lancet Oncol. 23, 91–103.

Jain, R.K., 2005. Normalization of tumor vasculature: an emerging concept in antiangiogenic therapy. Science 307, 58–62.

June, C.H., Sadelain, M., 2018. Chimeric antigen receptor therapy. N. Engl. J. Med. 379, 64–73.

June, C.H., O'Connor, R.S., Kawalekar, O.U., Ghassemi, S., Milone, M.C., 2018. CAR T cell immunotherapy for human cancer. Science 359, 1361–1365.

Kakarla, S., Chow, K.K., Mata, M., Shaffer, D.R., Song, X.T., Wu, M.F., Liu, H., Wang, L.L., Rowley, D.R., Pfizenmaier, K., Gottschalk, S., 2013. Antitumor effects of chimeric receptor engineered human T cells directed to tumor stroma. Mol. Ther. 21, 1611–1620.

Kieffer, Y., Hocine, H.R., Gentric, G., Pelon, F., Bernard, C., Bourachot, B., Lameiras, S., Albergante, L., Bonneau, C., Guyard, A., et al., 2020. Single-cell analysis reveals fibroblast clusters linked to immunotherapy resistance in cancer. Cancer Discov. 10, 1330–1351.

Kitamura, T., Qian, B.Z., Pollard, J.W., 2015. Immune cell promotion of metastasis. Nat. Rev. Immunol. 15, 73–86.

Kloss, C.C., Lee, J., Zhang, A., Chen, F., Melenhorst, J.J., Lacey, S.F., Maus, M.V., Fraietta, J.A., Zhao, Y., June, C.H., 2018. Dominant-negative TGF-beta receptor enhances PSMA-targeted human CAR T cell proliferation and augments prostate cancer eradication. Mol. Ther. 26, 1855–1866.

Kryczek, I., Zou, L., Rodriguez, P., Zhu, G., Wei, S., Mottram, P., Brumlik, M., Cheng, P., Curiel, T., Myers, L., et al., 2006. B7-H4 expression identifies a novel suppressive macrophage population in human ovarian carcinoma. J. Exp. Med. 203, 871–881.

Kuang, D.M., Zhao, Q., Peng, C., Xu, J., Zhang, J.P., Wu, C., Zheng, L., 2009. Activated monocytes in peritumoral stroma of hepatocellular carcinoma foster immune privilege and disease progression through PD-L1. J. Exp. Med. 206, 1327–1337.

Kuang, D.M., Peng, C., Zhao, Q., Wu, Y., Zhu, L.Y., Wang, J., Yin, X.Y., Li, L., Zheng, L., 2010. Tumor-activated monocytes promote expansion of IL-17-producing CD8+ T cells in hepatocellular carcinoma patients. J. Immunol. 185, 1544–1549.

Kumar, V., Donthireddy, L., Marvel, D., Condamine, T., Wang, F., Lavilla-Alonso, S., Hashimoto, A., Vonteddu, P., Behera, R., Goins, M.A., et al., 2017. Cancer-associated fibroblasts neutralize the anti-tumor effect of CSF1 receptor blockade by inducing PMN-MDSC infiltration of tumors. Cancer Cell 32, 654–668. e655.

Lambrechts, D., Wauters, E., Boeckx, B., Aibar, S., Nittner, D., Burton, O., Bassez, A., Decaluwe, H., Pircher, A., Van den Eynde, K., et al., 2018. Phenotype molding of stromal cells in the lung tumor microenvironment. Nat. Med. 24, 1277–1289.

Lanitis, E., Irving, M., Coukos, G., 2015. Targeting the tumor vasculature to enhance T cell activity. Curr. Opin. Immunol. 33, 55–63.

Lanitis, E., Kosti, P., Ronet, C., Cribioli, E., Rota, G., Spill, A., Reichenbach, P., Zoete, V., Dangaj Laniti, D., Coukos, G., Irving, M., 2021. VEGFR-2 redirected CAR-T cells are functionally impaired by soluble VEGF-A competition for receptor binding. J. Immunother. Cancer 9.

Li, J., Lee, Y., Li, Y., Jiang, Y., Lu, H., Zang, W., Zhao, X., Liu, L., Chen, Y., Tan, H., et al., 2018. Co-inhibitory molecule B7 superfamily member 1 expressed by tumor-infiltrating myeloid cells induces dysfunction of anti-tumor CD8(+) T cells. Immunity 48, 773–786. e775.

Li, H., Xiao, Y., Li, Q., Yao, J., Yuan, X., Zhang, Y., Yin, X., Saito, Y., Fan, H., Li, P., et al., 2022. The allergy mediator histamine confers resistance to immunotherapy in cancer patients via activation of the macrophage histamine receptor H1. Cancer Cell 40, 36–52. e39.

Lin, H., Wei, S., Hurt, E.M., Green, M.D., Zhao, L., Vatan, L., Szeliga, W., Herbst, R., Harms, P.W., Fecher, L.A., et al., 2018. Host expression of PD-L1 determines efficacy of PD-L1 pathway blockade-mediated tumor regression. J. Clin. Invest. 128, 805–815.

Lo, A., Wang, L.S., Scholler, J., Monslow, J., Avery, D., Newick, K., O'Brien, S., Evans, R.A., Bajor, D.J., Clendenin, C., et al., 2015. Tumor-promoting desmoplasia is disrupted by depleting FAP-expressing stromal cells. Cancer Res. 75, 2800–2810.

Lopez-Bujanda, Z., Drake, C.G., 2017. Myeloid-derived cells in prostate cancer progression: phenotype and prospective therapies. J. Leukoc. Biol. 102, 393–406.

Loyher, P.L., Hamon, P., Laviron, M., Meghraoui-Kheddar, A., Goncalves, E., Deng, Z., Torstensson, S., Bercovici, N., Baudesson de Chanville, C., Combadiere, B., et al., 2018. Macrophages of distinct origins contribute to tumor development in the lung. J. Exp. Med. 215, 2536–2553.

Luksza, M., Riaz, N., Makarov, V., Balachandran, V.P., Hellmann, M.D., Solovyov, A., Rizvi, N.A., Merghoub, T., Levine, A.J., Chan, T.A., et al., 2017. A neoantigen fitness model predicts tumour response to checkpoint blockade immunotherapy. Nature 551, 517–520.

Mantovani, A., Marchesi, F., Malesci, A., Laghi, L., Allavena, P., 2017. Tumour-associated macrophages as treatment targets in oncology. Nat. Rev. Clin. Oncol. 14, 399–416.

Mariathasan, S., Turley, S.J., Nickles, D., Castiglioni, A., Yuen, K., Wang, Y., Kadel III, E.E., Koeppen, H., Astarita, J.L., Cubas, R., et al., 2018. TGFbeta attenuates tumour response to PD-L1 blockade by contributing to exclusion of T cells. Nature 554, 544–548.

Maude, S.L., Laetsch, T.W., Buechner, J., Rives, S., Boyer, M., Bittencourt, H., Bader, P., Verneris, M.R., Stefanski, H.E., Myers, G.D., et al., 2018. Tisagenlecleucel in children and young adults with B-cell lymphoblastic leukemia. N. Engl. J. Med. 378, 439–448.

Mazanet, M.M., Hughes, C.C., 2002. B7-H1 is expressed by human endothelial cells and suppresses T cell cytokine synthesis. J. Immunol. 169, 3581–3588.

Medrek, C., Ponten, F., Jirstrom, K., Leandersson, K., 2012. The presence of tumor associated macrophages in tumor stroma as a prognostic marker for breast cancer patients. BMC Cancer 12, 306.

Mehta, A.K., Cheney, E.M., Hartl, C.A., Pantelidou, C., Oliwa, M., Castrillon, J.A., Lin, J.R., Hurst, K.E., de Oliveira Taveira, M., Johnson, N.T., et al., 2021. Targeting immunosuppressive macrophages overcomes PARP inhibitor resistance in BRCA1-associated triple-negative breast cancer. Nat. Cancer 2, 66–82.

Morello, A., Sadelain, M., Adusumilli, P.S., 2016. Mesothelin-targeted CARs: driving T cells to solid tumors. Cancer Discov. 6, 133–146.

Mount, C.W., Majzner, R.G., Sundaresh, S., Arnold, E.P., Kadapakkam, M., Haile, S., Labanieh, L., Hulleman, E., Woo, P.J., Rietberg, S.P., et al., 2018. Potent antitumor efficacy of anti-GD2 CAR T cells in H3-K27M(+) diffuse midline gliomas. Nat. Med. 24, 572–579.

Movahedi, K., Laoui, D., Gysemans, C., Baeten, M., Stange, G., Van den Bossche, J., Mack, M., Pipeleers, D., In't Veld, P., De Baetselier, P., Van Ginderachter, J.A., 2010. Different tumor microenvironments contain functionally distinct subsets of macrophages derived from Ly6C(high) monocytes. Cancer Res. 70, 5728–5739.

Munshi, N.C., Anderson Jr., L.D., Shah, N., Madduri, D., Berdeja, J., Lonial, S., Raje, N., Lin, Y., Siegel, D., Oriol, A., et al., 2021. Idecabtagene vicleucel in relapsed and refractory multiple myeloma. N. Engl. J. Med. 384, 705–716.

Nagarsheth, N., Wicha, M.S., Zou, W., 2017. Chemokines in the cancer microenvironment and their relevance in cancer immunotherapy. Nat. Rev. Immunol. 17, 559–572.

Nalawade, S.A., Shafer, P., Bajgain, P., McKenna, M.K., Ali, A., Kelly, L., Joubert, J., Gottschalk, S., Watanabe, N., Leen, A., et al., 2021. Selectively targeting myeloid-derived suppressor cells through TRAIL receptor 2 to enhance the efficacy of CAR T cell therapy for treatment of breast cancer. J. Immunother. Cancer 9.

Neelapu, S.S., Locke, F.L., Bartlett, N.L., Lekakis, L.J., Miklos, D.B., Jacobson, C.A., Braunschweig, I., Oluwole, O.O., Siddiqi, T., Lin, Y., et al., 2017. Axicabtagene ciloleucel CAR T-cell therapy in refractory large B-cell lymphoma. N. Engl. J. Med. 377, 2531–2544.

Neubert, N.J., Schmittnaegel, M., Bordry, N., Nassiri, S., Wald, N., Martignier, C., Tille, L., Homicsko, K., Damsky, W., Maby-El Hajjami, H., et al., 2018. T cell-induced CSF1 promotes melanoma resistance to PD1 blockade. Sci. Transl. Med. 10.

Noy, R., Pollard, J.W., 2014. Tumor-associated macrophages: from mechanisms to therapy. Immunity 41, 49–61.

O'Connell, J.T., Sugimoto, H., Cooke, V.G., MacDonald, B.A., Mehta, A.I., LeBleu, V.S., Dewar, R., Rocha, R.M., Brentani, R.R., Resnick, M.B., et al., 2011. VEGF-A and tenascin-C produced by S100A4+ stromal cells are important for metastatic colonization. Proc. Natl. Acad. Sci. U. S. A. 108, 16002–16007.

Ong, S.M., Tan, Y.C., Beretta, O., Jiang, D., Yeap, W.H., Tai, J.J., Wong, W.C., Yang, H., Schwarz, H., Lim, K.H., et al., 2012. Macrophages in human colorectal cancer are pro-inflammatory and prime T cells towards an anti-tumour type-1 inflammatory response. Eur. J. Immunol. 42, 89–100.

O'Shannessy, D.J., Somers, E.B., Wang, L.C., Wang, H., Hsu, R., 2015. Expression of folate receptors alpha and beta in normal and cancerous gynecologic tissues: correlation of expression of the beta isoform with macrophage markers. J. Ovarian Res. 8, 29.

Ozga, A.J., Chow, M.T., Luster, A.D., 2021. Chemokines and the immune response to cancer. Immunity 54, 859–874.

Parihar, R., Rivas, C., Huynh, M., Omer, B., Lapteva, N., Metelitsa, L.S., Gottschalk, S.M., Rooney, C.M., 2019. NK cells expressing a chimeric activating receptor eliminate MDSCs and rescue impaired CAR-T cell activity against solid tumors. Cancer Immunol. Res. 7, 363–375.

Pasquini, M.C., Hu, Z.H., Curran, K., Laetsch, T., Locke, F., Rouce, R., Pulsipher, M.A., Phillips, C.L., Keating, A., Frigault, M.J., et al., 2020. Real-world evidence of tisagenlecleucel for pediatric acute lymphoblastic leukemia and non-Hodgkin lymphoma. Blood Adv. 4, 5414–5424.

Petrausch, U., Schuberth, P.C., Hagedorn, C., Soltermann, A., Tomaszek, S., Stahel, R., Weder, W., Renner, C., 2012. Re-directed T cells for the treatment of fibroblast activation protein (FAP)-positive malignant pleural mesothelioma (FAPME-1). BMC Cancer 12, 615.

Petrovic, K., Robinson, J., Whitworth, K., Jinks, E., Shaaban, A., Lee, S.P., 2019. TEM8/ANTXR1-specific CAR T cells mediate toxicity in vivo. PLoS One 14, e0224015.

Pircher, M., Schuberth, P., Gulati, P., Sulser, S., Weder, W., Curioni, A., Renner, C., Petrausch, U., 2015. FAP-specific re-directed T cells first in-man study in malignant pleural mesothelioma: experience of the first patient treated. J. Immunother. Cancer 3, P120.

Pituch, K.C., Miska, J., Krenciute, G., Panek, W.K., Li, G., Rodriguez-Cruz, T., Wu, M., Han, Y., Lesniak, M.S., Gottschalk, S., Balyasnikova, I.V., 2018. Adoptive transfer of IL13Ralpha2-specific chimeric antigen receptor T cells creates a pro-inflammatory environment in glioblastoma. Mol. Ther. 26, 986–995.

Pollard, J.W., 2004. Tumour-educated macrophages promote tumour progression and metastasis. Nat. Rev. Cancer 4, 71–78.

Pothoff, G., Curtius, J.M., Wassermann, K., Junge-Hulsing, M., Sechtem, U., Schicha, H., Hilger, H.H., 1992. Transesophageal echography in staging of bronchial cancers. Pneumologie 46, 111–117.

Propper, D.J., Balkwill, F.R., 2022. Harnessing cytokines and chemokines for cancer therapy. Nat. Rev. Clin. Oncol.

Puig-Kroger, A., Sierra-Filardi, E., Dominguez-Soto, A., Samaniego, R., Corcuera, M.T., Gomez-Aguado, F., Ratnam, M., Sanchez-Mateos, P., Corbi, A.L., 2009. Folate receptor beta is expressed by tumor-associated macrophages and constitutes a marker for M2 anti-inflammatory/regulatory macrophages. Cancer Res. 69, 9395–9403.

Qian, B.Z., Pollard, J.W., 2010. Macrophage diversity enhances tumor progression and metastasis. Cell 141, 39–51.

Qian, B.Z., Li, J., Zhang, H., Kitamura, T., Zhang, J., Campion, L.R., Kaiser, E.A., Snyder, L.A., Pollard, J.W., 2011. CCL2 recruits inflammatory monocytes to facilitate breast-tumour metastasis. Nature 475, 222–225.

Rafiq, S., Hackett, C.S., Brentjens, R.J., 2020. Engineering strategies to overcome the current roadblocks in CAR T cell therapy. Nat. Rev. Clin. Oncol. 17, 147–167.

Riabov, V., Gudima, A., Wang, N., Mickley, A., Orekhov, A., Kzhyshkowska, J., 2014. Role of tumor associated macrophages in tumor angiogenesis and lymphangiogenesis. Front. Physiol. 5, 75.

Ries, C.H., Cannarile, M.A., Hoves, S., Benz, J., Wartha, K., Runza, V., Rey-Giraud, F., Pradel, L.P., Feuerhake, F., Klaman, I., et al., 2014. Targeting tumor-associated macrophages with anti-CSF-1R antibody reveals a strategy for cancer therapy. Cancer Cell 25, 846–859.

Rizvi, N.A., Hellmann, M.D., Snyder, A., Kvistborg, P., Makarov, V., Havel, J.J., Lee, W., Yuan, J., Wong, P., Ho, T.S., et al., 2015. Cancer immunology. Mutational landscape determines sensitivity to PD-1 blockade in non-small cell lung cancer. Science 348, 124–128.

Rodriguez, E., Boelaars, K., Brown, K., Eveline Li, R.J., Kruijssen, L., Bruijns, S.C.M., van Ee, T., Schetters, S.T.T., Crommentuijn, M.H.W., van der Horst, J.C., et al., 2021. Sialic acids in pancreatic cancer cells drive tumour-associated macrophage differentiation via the Siglec receptors Siglec-7 and Siglec-9. Nat. Commun. 12, 1270.

Rodriguez-Garcia, A., Palazon, A., Noguera-Ortega, E., Powell Jr., D.J., Guedan, S., 2020. CAR-T cells hit the tumor microenvironment: strategies to overcome tumor escape. Front. Immunol. 11, 1109.

Rodriguez-Garcia, A., Lynn, R.C., Poussin, M., Eiva, M.A., Shaw, L.C., O'Connor, R.S., Minutolo, N.G., Casado-Medrano, V., Lopez, G., Matsuyama, T., Powell Jr., D.J., 2021. CAR-T cell-mediated depletion of immunosuppressive tumor-associated macrophages promotes endogenous antitumor immunity and augments adoptive immunotherapy. Nat. Commun. 12, 877.

Ruella, M., Klichinsky, M., Kenderian, S.S., Shestova, O., Ziober, A., Kraft, D.O., Feldman, M., Wasik, M.A., June, C.H., Gill, S., 2017. Overcoming the immunosuppressive tumor microenvironment of Hodgkin lymphoma using chimeric antigen receptor T cells. Cancer Discov. 7, 1154–1167.

Ruffell, B., Chang-Strachan, D., Chan, V., Rosenbusch, A., Ho, C.M., Pryer, N., Daniel, D., Hwang, E.S., Rugo, H.S., Coussens, L.M., 2014. Macrophage IL-10 blocks CD8+ T cell-dependent responses to chemotherapy by suppressing IL-12 expression in intratumoral dendritic cells. Cancer Cell 26, 623–637.

Ryner, L., Guan, Y., Firestein, R., Xiao, Y., Choi, Y., Rabe, C., Lu, S., Fuentes, E., Huw, L.Y., Lackner, M.R., et al., 2015. Upregulation of Periostin and reactive stroma is associated with primary chemoresistance and predicts clinical outcomes in epithelial ovarian cancer. Clin. Cancer Res. 21, 2941–2951.

Sadelain, M., Brentjens, R., Riviere, I., 2013. The basic principles of chimeric antigen receptor design. Cancer Discov. 3, 388–398.
Sahai, E., Astsaturov, I., Cukierman, E., DeNardo, D.G., Egeblad, M., Evans, R.M., Fearon, D., Greten, F.R., Hingorani, S.R., Hunter, T., et al., 2020. A framework for advancing our understanding of cancer-associated fibroblasts. Nat. Rev. Cancer 20, 174–186.
Salvagno, C., Ciampricotti, M., Tuit, S., Hau, C.S., van Weverwijk, A., Coffelt, S.B., Kersten, K., Vrijland, K., Kos, K., Ulas, T., et al., 2019. Therapeutic targeting of macrophages enhances chemotherapy efficacy by unleashing type I interferon response. Nat. Cell Biol. 21, 511–521.
Santoro, S.P., Kim, S., Motz, G.T., Alatzoglou, D., Li, C., Irving, M., Powell Jr., D.J., Coukos, G., 2015. T cells bearing a chimeric antigen receptor against prostate-specific membrane antigen mediate vascular disruption and result in tumor regression. Cancer Immunol. Res. 3, 68–84.
Schuster, S.J., Investigators, J., 2019. Tisagenlecleucel in diffuse large B-cell lymphoma. Reply. N. Engl. J. Med. 380, 1586.
Sharma, V.P., Tang, B., Wang, Y., Duran, C.L., Karagiannis, G.S., Xue, E.A., Entenberg, D., Borriello, L., Coste, A., Eddy, R.J., et al., 2021. Live tumor imaging shows macrophage induction and TMEM-mediated enrichment of cancer stem cells during metastatic dissemination. Nat. Commun. 12, 7300.
Shen, J., Putt, K.S., Visscher, D.W., Murphy, L., Cohen, C., Singhal, S., Sandusky, G., Feng, Y., Dimitrov, D.S., Low, P.S., 2015. Assessment of folate receptor-beta expression in human neoplastic tissues. Oncotarget 6, 14700–14709.
Smith, L.K., Boukhaled, G.M., Condotta, S.A., Mazouz, S., Guthmiller, J.J., Vijay, R., Butler, N.S., Bruneau, J., Shoukry, N.H., Krawczyk, C.M., Richer, M.J., 2018. Interleukin-10 directly inhibits CD8(+) T cell function by enhancing N-glycan branching to decrease antigen sensitivity. Immunity 48, 299–312. e295.
Steele, C.W., Karim, S.A., Leach, J.D.G., Bailey, P., Upstill-Goddard, R., Rishi, L., Foth, M., Bryson, S., McDaid, K., Wilson, Z., et al., 2016. CXCR2 inhibition profoundly suppresses metastases and augments immunotherapy in pancreatic ductal adenocarcinoma. Cancer Cell 29, 832–845.
Steidl, C., Lee, T., Shah, S.P., Farinha, P., Han, G., Nayar, T., Delaney, A., Jones, S.J., Iqbal, J., Weisenburger, D.D., et al., 2010. Tumor-associated macrophages and survival in classic Hodgkin's lymphoma. N. Engl. J. Med. 362, 875–885.
Stromnes, I.M., Burrack, A.L., Hulbert, A., Bonson, P., Black, C., Brockenbrough, J.S., Raynor, J.F., Spartz, E.J., Pierce, R.H., Greenberg, P.D., Hingorani, S.R., 2019. Differential effects of depleting versus programming tumor-associated macrophages on engineered T cells in pancreatic ductal adenocarcinoma. Cancer Immunol. Res. 7, 977–989.
Su, S., Liu, Q., Chen, J., Chen, J., Chen, F., He, C., Huang, D., Wu, W., Lin, L., Huang, W., et al., 2014. A positive feedback loop between mesenchymal-like cancer cells and macrophages is essential to breast cancer metastasis. Cancer Cell 25, 605–620.
Szekely, B., Bossuyt, V., Li, X., Wali, V.B., Patwardhan, G.A., Frederick, C., Silber, A., Park, T., Harigopal, M., Pelekanou, V., et al., 2018. Immunological differences between primary and metastatic breast cancer. Ann. Oncol. 29, 2232–2239.
Tang, H., Wang, Y., Chlewicki, L.K., Zhang, Y., Guo, J., Liang, W., Wang, J., Wang, X., Fu, Y.X., 2016. Facilitating T cell infiltration in tumor microenvironment overcomes resistance to PD-L1 blockade. Cancer Cell 29, 285–296.
Tauriello, D.V.F., Palomo-Ponce, S., Stork, D., Berenguer-Llergo, A., Badia-Ramentol, J., Iglesias, M., Sevillano, M., Ibiza, S., Canellas, A., Hernando-Momblona, X., et al., 2018. TGFbeta drives immune evasion in genetically reconstituted colon cancer metastasis. Nature 554, 538–543.

Theruvath, J., Sotillo, E., Mount, C.W., Graef, C.M., Delaidelli, A., Heitzeneder, S., Labanieh, L., Dhingra, S., Leruste, A., Majzner, R.G., et al., 2020. Locoregionally administered B7-H3-targeted CAR T cells for treatment of atypical teratoid/rhabdoid tumors. Nat. Med. 26, 712–719.

Torroella-Kouri, M., Silvera, R., Rodriguez, D., Caso, R., Shatry, A., Opiela, S., Ilkovitch, D., Schwendener, R.A., Iragavarapu-Charyulu, V., Cardentey, Y., et al., 2009. Identification of a subpopulation of macrophages in mammary tumor-bearing mice that are neither M1 nor M2 and are less differentiated. Cancer Res. 69, 4800–4809.

Tran, E., Chinnasamy, D., Yu, Z., Morgan, R.A., Lee, C.C., Restifo, N.P., Rosenberg, S.A., 2013. Immune targeting of fibroblast activation protein triggers recognition of multipotent bone marrow stromal cells and cachexia. J. Exp. Med. 210, 1125–1135.

Valkenburg, K.C., de Groot, A.E., Pienta, K.J., 2018. Targeting the tumour stroma to improve cancer therapy. Nat. Rev. Clin. Oncol. 15, 366–381.

VanSeggelen, H., Hammill, J.A., Dvorkin-Gheva, A., Tantalo, D.G., Kwiecien, J.M., Denisova, G.F., Rabinovich, B., Wan, Y., Bramson, J.L., 2015. T cells engineered with chimeric antigen receptors targeting NKG2D ligands display lethal toxicity in mice. Mol. Ther. 23, 1600–1610.

Veglia, F., Sanseviero, E., Gabrilovich, D.I., 2021. Myeloid-derived suppressor cells in the era of increasing myeloid cell diversity. Nat. Rev. Immunol. 21, 485–498.

Vennin, C., Melenec, P., Rouet, R., Nobis, M., Cazet, A.S., Murphy, K.J., Herrmann, D., Reed, D.A., Lucas, M.C., Warren, S.C., et al., 2019. CAF hierarchy driven by pancreatic cancer cell p53-status creates a pro-metastatic and chemoresistant environment via perlecan. Nat. Commun. 10, 3637.

Vitanza, N.A., Johnson, A.J., Wilson, A.L., Brown, C., Yokoyama, J.K., Kunkele, A., Chang, C.A., Rawlings-Rhea, S., Huang, W., Seidel, K., et al., 2021. Locoregional infusion of HER2-specific CAR T cells in children and young adults with recurrent or refractory CNS tumors: an interim analysis. Nat. Med. 27, 1544–1552.

Wan, S., Zhao, E., Kryczek, I., Vatan, L., Sadovskaya, A., Ludema, G., Simeone, D.M., Zou, W., Welling, T.H., 2014. Tumor-associated macrophages produce interleukin 6 and signal via STAT3 to promote expansion of human hepatocellular carcinoma stem cells. Gastroenterology 147, 1393–1404.

Wan, S., Kuo, N., Kryczek, I., Zou, W., Welling, T.H., 2015. Myeloid cells in hepatocellular carcinoma. Hepatology 62, 1304–1312.

Wang, W., Ma, Y., Li, J., Shi, H.S., Wang, L.Q., Guo, F.C., Zhang, J., Li, D., Mo, B.H., Wen, F., et al., 2013. Specificity redirection by CAR with human VEGFR-1 affinity endows T lymphocytes with tumor-killing ability and anti-angiogenic potency. Gene Ther. 20, 970–978.

Wang, L.C., Lo, A., Scholler, J., Sun, J., Majumdar, R.S., Kapoor, V., Antzis, M., Cotner, C.E., Johnson, L.A., Durham, A.C., et al., 2014. Targeting fibroblast activation protein in tumor stroma with chimeric antigen receptor T cells can inhibit tumor growth and augment host immunity without severe toxicity. Cancer Immunol. Res. 2, 154–166.

Wang, W., Kryczek, I., Dostal, L., Lin, H., Tan, L., Zhao, L., Lu, F., Wei, S., Maj, T., Peng, D., et al., 2016. Effector T cells abrogate stroma-mediated chemoresistance in ovarian cancer. Cell 165, 1092–1105.

Wang, J., Sun, J., Liu, L.N., Flies, D.B., Nie, X., Toki, M., Zhang, J., Song, C., Zarr, M., Zhou, X., et al., 2019. Siglec-15 as an immune suppressor and potential target for normalization cancer immunotherapy. Nat. Med. 25, 656–666.

Wang, M., Munoz, J., Goy, A., Locke, F.L., Jacobson, C.A., Hill, B.T., Timmerman, J.M., Holmes, H., Jaglowski, S., Flinn, I.W., et al., 2020. KTE-X19 CAR T-cell therapy in relapsed or refractory mantle-cell lymphoma. N. Engl. J. Med. 382, 1331–1342.

Williams, C.B., Yeh, E.S., Soloff, A.C., 2016. Tumor-associated macrophages: unwitting accomplices in breast cancer malignancy. NPJ Breast Cancer 2.

Wu, Q., Zhou, W., Yin, S., Zhou, Y., Chen, T., Qian, J., Su, R., Hong, L., Lu, H., Zhang, F., et al., 2019. Blocking triggering receptor expressed on myeloid Cells-1-positive tumor-associated macrophages induced by hypoxia reverses immunosuppression and anti-programmed cell death ligand 1 resistance in liver cancer. Hepatology 70, 198–214.

Xie, Y.J., Dougan, M., Jailkhani, N., Ingram, J., Fang, T., Kummer, L., Momin, N., Pishesha, N., Rickelt, S., Hynes, R.O., Ploegh, H., 2019. Nanobody-based CAR T cells that target the tumor microenvironment inhibit the growth of solid tumors in immunocompetent mice. Proc. Natl. Acad. Sci. U. S. A. 116, 7624–7631.

Xin, L., Gao, J., Zheng, Z., Chen, Y., Lv, S., Zhao, Z., Yu, C., Yang, X., Zhang, R., 2021. Fibroblast activation protein-alpha as a target in the bench-to-bedside diagnosis and treatment of tumors: a narrative review. Front. Oncol. 11, 648187.

Yamashita, M., Ogawa, T., Zhang, X., Hanamura, N., Kashikura, Y., Takamura, M., Yoneda, M., Shiraishi, T., 2012. Role of stromal myofibroblasts in invasive breast cancer: stromal expression of alpha-smooth muscle actin correlates with worse clinical outcome. Breast Cancer 19, 170–176.

Yin, Y., Yao, S., Hu, Y., Feng, Y., Li, M., Bian, Z., Zhang, J., Qin, Y., Qi, X., Zhou, L., et al., 2017. The immune-microenvironment confers chemoresistance of colorectal cancer through macrophage-derived IL6. Clin. Cancer Res. 23, 7375–7387.

Yona, S., Kim, K.W., Wolf, Y., Mildner, A., Varol, D., Breker, M., Strauss-Ayali, D., Viukov, S., Guilliams, M., Misharin, A., et al., 2013. Fate mapping reveals origins and dynamics of monocytes and tissue macrophages under homeostasis. Immunity 38, 79–91.

Yu, Y., Xiao, C.H., Tan, L.D., Wang, Q.S., Li, X.Q., Feng, Y.M., 2014. Cancer-associated fibroblasts induce epithelial-mesenchymal transition of breast cancer cells through paracrine TGF-beta signalling. Br. J. Cancer 110, 724–732.

Zaretsky, J.M., Garcia-Diaz, A., Shin, D.S., Escuin-Ordinas, H., Hugo, W., Hu-Lieskovan, S., Torrejon, D.Y., Abril-Rodriguez, G., Sandoval, S., Barthly, L., et al., 2016. Mutations associated with acquired resistance to PD-1 blockade in melanoma. N. Engl. J. Med. 375, 819–829.

Zhang, T., Sentman, C.L., 2013. Mouse tumor vasculature expresses NKG2D ligands and can be targeted by chimeric NKG2D-modified T cells. J. Immunol. 190, 2455–2463.

Zhu, Y., Herndon, J.M., Sojka, D.K., Kim, K.W., Knolhoff, B.L., Zuo, C., Cullinan, D.R., Luo, J., Bearden, A.R., Lavine, K.J., et al., 2017. Tissue-resident macrophages in pancreatic ductal adenocarcinoma originate from embryonic hematopoiesis and promote tumor progression. Immunity 47, 323–338. e326.

Zhu, L., Narloch, J.L., Onkar, S., Joy, M., Broadwater, G., Luedke, C., Hall, A., Kim, R., Pogue-Geile, K., Sammons, S., et al., 2019. Metastatic breast cancers have reduced immune cell recruitment but harbor increased macrophages relative to their matched primary tumors. J. Immunother. Cancer 7, 265.

CHAPTER FIVE

What will (and should) be improved in CAR immunotherapy?

Europa Azucena González-Navarro, Marta Español, Natalia Egri, Maria Castellà, Hugo Calderón, Carolina España, Carla Guijarro, Libertad Heredia, Mariona Pascal, and Manel Juan Otero*

Servei d'Immunologia. Hospital Clínic de Barcelona. IDIBAPS. Universitat de Barcelona. Plataforma d'Immunoteràpia Hospital Clínic—Hospital Sant Joan de Déu, Barcelona, Spain
*Corresponding author: e-mail address: mjuan@clinic.cat

Contents

1. General concepts	150
2. Detection of Targets: Specificity and Affinity	151
3. Global molecular structure: From signaling to combination	153
4. Gene transfer protocols: Vectors, gene editing and upper CAR generation	154
5. Cell involvement: Allogenic, autologous and cell populations	155
6. Clinical protocols for personalized use of the product	157
7. Combination of therapies and procedures	158
8. Regulatory improvements	158
9. Pharma and Academic collaboration for supplying sustainable options to the health systems	159
10. Final discussion and conclusion	159
Acknowledgments	160
References	160

Abstract

Chimeric antigen receptor (CAR) is probably one of the most successful proposals for cancer treatment, especially hematological diseases for which several Advanced Therapies Medicinal Products (ATMP) have been approved worldwide by drug agencies. But, despite this unprecedented success in the oncology and cell/gene therapy fields, there are a lot of aspects that could (and should) be improved in the multiple aspects that involve this complex therapy: from the design of the chimeric molecule to the clinical protocols of use of the engineered T-cells, including even the regulatory rules that they are currently restricting the development of these hopeful therapies. In this chapter, we will try to summarize the main aspects that can (and probably should) be improved for the expansion of immunotherapy with CAR proposals beyond onco-hematology.

Abbreviations

2G	second generation
aSCT	allogenic Stem Cell Transplantation
ATMP	advanced therapy medicinal product
CAR	chimeric antigen receptor
CM	central memory
NK	natural killer
scFv	single chain variable regions
SCM	stem-cell memory
TcR	T cell Receptor
TF	transcription factors
TRAC	T cell receptor alpha constant
TRUCK	T cells redirected for universal cytokine killing
UCAR-T	Universal CAR-T
VHH	variable heavy-chain

1. General concepts

Before 2013 when the Science magazine deemed **cancer immunotherapy** the 2013 "Breakthrough of the Year" (Couzin-Frankel, 2013), next to the use of check point-inhibitors, the developing of chimeric antigen receptors (CAR) in engineered T-Cells (CAR-T) was already considered a clear achievement based, at that moment, in just few patients and initial descriptions (Brentjens et al., 2011; Kochenderfer et al., 2010; Porter et al., 2011). Nowadays, after more than 10,000 patients worldwide treated with CAR T-cell therapies (Burki, 2021), these "Advanced Therapies" (combining gene modifications with cell products) are real tools for cancer treatment that could and should make a step forward beyond oncohematology.

Six are the current CAR-T therapies approved by drug agencies in the world (Kymriah®, Yescarta®, Tecartus®, Breyanzi® and ARI-0001 for CAR-T against CD19+ B-cell lymphoproliferative diseases and Abecma® for CAR-T anti-BCMA in multiple myeloma), but there are tens of new proposals under evaluation at the clinical trial level.

First of all we should understand that although CAR therapy for the regulatory point of view is globally considered a "drug," it is important to have in mind that it is more near to a clinical personalized protocol or tool for improving a internal function (immune response), that to a conventional chemical product externally prepared in a homogenous way to be used

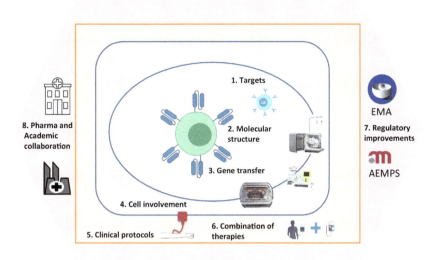

Fig. 1 8 key challenges to improve CAR therapies.

for a specific disease. Although CARs is an acronym defining molecules, CAR therapy involve these molecules, the engineered cells, the production protocol, the different clinical protocols and even the rules that define the possibilities of use. In this therapeutic *tool* there are at least eight key challenges where it is possible to improve and personalize the treatment (Fig. 1):

1. Detection of targets: specificity and affinity.
2. Global molecular structure: From signaling to combination.
3. Gene transfer protocols: Vectors, gene editing and upper CAR generation.
4. Cell involvement: Allogenic, autologous and cell populations.
5. Clinical protocols for personalized use of the product.
6. Combination of therapies and procedures.
7. Regulatory improvements.
8. Pharma and Academic collaboration for supplying sustainable options to the health systems.

2. Detection of Targets: Specificity and Affinity

As any receptor, CAR molecules direct the cells where they are sited to recognize the target molecule being the specificity and affinity the main aspects that can determine unexpected events and the selectivity of function.

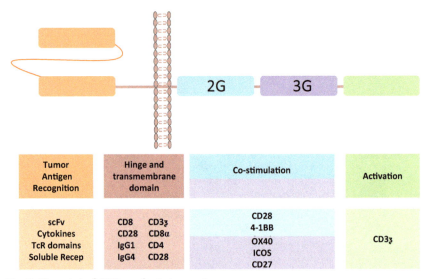

Fig. 2 Blueprint of CAR products improving.

When the **recognizing domain** is obtained from antibodies (characteristically from single chain variable regions from antibodies, scFv), this specificity can be very selective while a high strength of this recognition (affinity) can produce cross-reactivities to other similar molecules with still enough binding strength for producing unexpected effect. Although other binders (cytokines, nanobodies soluble receptors, TcR domains, …) are being used (Fig. 2), the main source of receptors is by now scFv, moving from murine domains to most humanized structures to reduce immunogenicity. Variable fragment of heavy-chain antibodies of *Cameliadae* (VHH) composed of only two heavy chains (with no light chains) are a very promise alternative due to their smaller in size that facilitates the generation of CAR with the potentiality of several domains that can easily multiply the recognizable targets. In addition, this approach being smaller can generate less immunogenicity and, at the same time, it seems to decrease the tonic activation of T-cells (Han et al., 2021). scFv or VHH are relevant elements that could define the persistence of the product once it is introduced in the patient. While the use of scFv reduces the spectrum of target antigen to surface antigens, the option of domains derived of TcR (T cell Receptor) could allow the recognition of intracellular antigens, although often this approximation could reduce the specificity. Other element related with targets are the **linkers** used to the expression of complementary domains. Linkers are helpful to proper folding the scFv so it is capable of antigen recognition and binding (Dwivedi et al., 2019).

They are, in general, flexible structures that can determine the interaction and add immunogenicity, ... similarly to the **hinge** regions (derived from CD8, CD28 or Fc IgG domains) used to link the binding domains to transmembrane and consequently to signaling sequences. On the other hand, differences in the length and composition of the hinge can affect antigen binding and signaling through the CAR (Fig. 2) (Jensen and Riddell, 2015).

3. Global molecular structure: From signaling to combination

Next to the relevance of the selectivity of the binding domains, signaling sequences of CAR are key elements for defining the final function of the product. Signaling domain, typically comprises an activation domain and one or more costimulatory domain. 2G (Second Generation) CARs with CD3zeta as activation domain next to a costimulatory domain (CD28 or 4-1BB) are the main used structures, 3G CARs (with a second costimulatory domain) are under study to increase the final effect of the construct (Fig. 2). Other proposals can introduce additional recognitions that can increase specificity (dual CARs) or block effect (inhibitory CARs). To avoid the loss of expression of tumor antigens in the neoplastic cells, different improvement approaches are being proposed, including combinations of co-expression of two (or more) complementary CARs. Another approach consists of the generation of CARs with different targets (biscistronic CARs), transducing a T-cell with two different lentiviruses at the same time. In the same way, the sequential administration of different monospecific CARs was also carried out. Other possibility is using CARs in parallel, in a juxta-membrane localization, in which two CARs are co-expressed across the cell membrane, one of which is linked to an activation domain and the other to a costimulatory molecule (Muliaditan et al., 2021). In 3G (or in other proposals) it is possible to find different co-stimulatory domains (such as ICOS or other domains) that in just one molecule or in complementary genes (4G CARs) can modulate the effect of the CAR-Ts. Sometimes these domains can define the expression of sequentially regulated genes or producing cytokines (the so-called TRUCK from "T-cells Redirected for Universal Cytokine Killing") that could modify the tumor microenvironment (Chmielewski et al., 2014). In fact, options of combinations become "almost infinite," and the main limitation is accomplished to demonstrate better efficacies of these proposals through clinical trials. In this way, when compared 2G with 3G designs, some 3G CARs elicit inferior anti-tumor activity (Hombach et al., 2013). One potential explanation is that the linear nature of the 3G CAR endodomain produces the ineffective positioning of one co-stimulatory module away from the membrane (Muliaditan et al., 2021), and impairing the downstream signaling. Finally, 4G proposals (in fact with just 2G instead of 3G CARs) could show a better anti-tumor cytotoxic function, being effective for the eradication of established tumors, even without a prior conditioning regimen (Pegram et al., 2012).

4. Gene transfer protocols: Vectors, gene editing and upper CAR generation

As in other gene therapies, transfer protocols of the modifying gene are major aspects to have in mind. The use of virus as a **vector** (especially lentivirus or retrovirus) is one of the main procedures used to engine T-cells, because gene integration allows that after the expected proliferation post-antigen recognition, "daughter cells" continue to express the CAR. Non-integrative virus (such adenovirus or adeno-associated virus) can be used as options for transient expression of genes in cells where the proliferation is not a key step (e.g., macrophages), being a way to control the expression of a transgene under study as a point of safety. **Promoters** in the vector are also important elements to have in mind and the site of integration should be also controlled to define final function of the transgene. In this sense, the integration of the gene under the control of TRAC (T cell Receptor Alpha Constant) promoter has demonstrated that improve the function of CAR in T-cells. The use of gene editing tools, trying to direct the insertion in the TRAC gene allow a "more physiological" expression of the transgene under TRAC promoter and the knock-out of the endogenous TCR that can reduce alloreactivity of these cells (Tristán-Manzano et al., 2021). Although the use of "older" proposals such TALENS o Zinc-finger editing have been also used, CRISPR-Cas9 gene editing is becoming the more frequent approach for multi-gene editing in T-cells, not only by reducing both-direction allogenicity (knocking-out TRAC and/or beta2-microglobuline for abrogating HLA expression) but also introducing new gene for additional modifications in 4G CARs (Eyquem et al., 2017). These approaches have the dual advantage of eliminating the endogenous TcR, while simultaneously enhancing CAR cell function by reducing tonic signaling (Rafiq et al., 2020). Gene editing procedures appears as the main option for the development of the usually named "Universal CAR-T," a "product off-the-shelf," ready to be use for patients, because it is previously produced from cells of one or several donors. This UCAR-T is now one of the main aims for the industry due to these CAR-T cells could be manufactured in bulk from healthy donors to be readily available for use in a timely manner and they would avoid the current personalized production typically used in most of the approved engineered cell products. Although the option of UCAR-T would open the option for a more homogeneous, ready to use and controlled product (more like other drugs for pharmaceutical companies), they are currently less effective than autologous products. Using non-viral vectors as transposon system can be another possibility, as a valid cost-effective option. DNA transposons stably integrate, through a precise recombinase-mediated mechanism, into chromosomes, providing long-term expression of the gene of interest (Magnani et al., 2020). The advantage of this gene delivery for CART cell therapy is a less expensive production, because no GMP-grade virus generation is needed.

5. Cell involvement: Allogenic, autologous and cell populations

Although we commented in the previous section that UCAR-T and, in general, all of off-the-shelf donor-derived engineered cells, could be the "Holy Grail" for industrial ATMP, currently all the commercially available CAR-T products and most of the proposal under clinical trial are engineered autologous T-cells from the own patients. Unexpectedly this personalized procedure is considered enough "homogeneous" for accepting as "equivalent products" by regulatory agencies despite getting form patients with or without allogeneic Stem Cell Transplantation (aSCT). Even if cells are clearly genetically different when are obtained after aSCT, for a regulatory point of view these "grafted cells" are considered equivalent to own cells (autologous products), while a CAR T cell obtained from cells freshly obtained from the same donor would be considered allogeneic and if they are never used as a source of the CAR-T, they will be evaluated as a different product. In general the source of cells, the type of selected cells and even the system of activation or culture are considered important variables for defining new CAR-T products. This is the case of Yescarta® or Tecartus®, where the system of selection and cell production define two different "drugs" even when the CAR molecule is the same.

A promising new technology that could be used instead of allogenic approach is the use of exosomes. Exosomes from HEK293T

CAR-expressing cells can be a delivery system of functional CAR particles that can be used as ATMP instead of conventional CAR-T cells. Exosomes expressing CAR in their surface can induce cytotoxicity when are put in contact with target cells. One of the clearer benefit of this approach is to cut short the step of ex-vivo cultures of CAR engineered T-cells (Haque and Vaiselbuh, 2021).

We cannot forget the intrinsic defects of T-cells, in the lack of effectiveness of CAR treatments, for which it becomes necessary to define new proposals to improve this T-cell function. For a biological point of view, the population and subpopulations used for engineering the CAR gne, is clearly a major aspect to define the different functions and efficacies of the cell-products (Fig. 3). Although all the commercially available products are a mixture of CD4+ and CD8+, in general the proportion of both is very variable. Selection of type and quantities of each population are key aspects under studies. Next to CD4 and CD8+ populations, the presence of Stem-Cell Memory (SCM) o Central Memory (CM) T-cell phenotypes seems to be crucial for the persistence of CAR-T. Despite these cells can be differentiated to effector T-cells, an initial proportion of this subpopulation is important to have a fast antitumor effect. T cells expressing CARs with CD28 domains differentiate into effector memory T cells, whereas 4-1BB-containing CAR T cells differentiate into central memory T cells (Kawalekar et al., 2016). Furthermore, it has been seen that some cells (named "stem cell-like cells") produce other more differentiated and effector cells retaining the function and total number of T-cells even after these

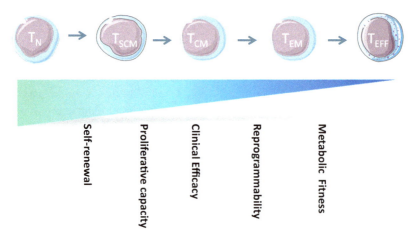

Fig. 3 Functions and efficacies according cell types used for development of CAR T cell immunotherapy.

effector cells disappear or become unfunctional by the presence of immune checkpoints. There are other co-stimulatory domains (such as OX40, ICOS or CD27) that can contribute to the functional variability of CARs but have not yet been tested in patients. It has been considered that these co-stimulatory domains can influence the exhaustion of CARs; some current therapeutic proposals try to avoid unfunctional T-cells (such as inhibition of CAR signaling) by enhancing CAR T-cell fitness, preventing T-cell exhaustion, e.g., blocking immune checkpoint inhibitors, knocking-down transcription factors (TF) upregulated in dysfunctional CAR T-cells, or upregulating those TF inductors of resistance to exhaustion. In the same way, epigenetic alterations are related with T-cell exhaustion with limited potential for reversion (Good et al., 2021; Lynn et al., 2019; Weber et al., 2021).

Similarly, to T-cells, NK (Natural Killer) cells are also a "more than promising" option for engineering with CAR (CAR-NK), because of their low alloreactivity. Completed clinical trial point out that this approach can be used as an "off-the-shelf" "ready-to-use" product (Liu et al., 2020) but has been associated with decreased persistence of the CAR-NK in patients (Rotolo et al., 2019). In one step behind, gamma-delta T-cells can appear as a cytotoxic option for anti-tumoral low-allogenic products due to the CAR and their antitumor TCRs (Sebestyen et al., 2020). Finally, Macrophages can be also modified with CAR (CAR-M) to improve and modulate their important role as tumor-associated macrophages (TAMs) toward anti-tumoral functions (Klichinsky et al., 2020).

6. Clinical protocols for personalized use of the product

Most of the CAR-T products are defined not only by the production but also by very specific and limiting protocols, including on-site manipulation (e.g., with some CAR-Ts, cryopreserved product before the infusion should be conserved in the Dewar for the transportation before the infusion, in other cases, the product can be transferred to conventional liquid nitrogen tanks waiting for infusion moment) or especially the infusion procedures. Although CAR-T infusion in a single dose is mainly used for most of the commercial, there are strong evidence that the fractionated increasing doses at infusion (e.g., 10%, 30% and 60%) control better the potential clinical adverse effects with similar efficacy. Unfortunately, although the reduction of adverse effects could be a clear advantage for patients, this change in the clinical management is so strictly defined that would require a different clinical trial to demonstrate that. In any case, this is just an example of different aspects related to clinical management of the protocol that could improve the result of CAR-T treatment. Conditioning protocol, time and dose of the infusion of immunoglobulins, use of tocilizumab, corticoids, anakinra ... can suppose improvement different aspects of the clinical management of the patients.

7. Combination of therapies and procedures

In general immunotherapy is being used as monotherapy, although it is clear that the combination of treatments can improve the final effect of each one. Next to combination of CAR-T with stimulating cytokines, check-point inhibitors can help to avoid exhaustion of the CAR-T. Although there are defined cases that suggest that these combinations can solve some limitations of the CAR-T therapy (Chong et al., 2017; Li et al., 2018), only additional clinical trials can really provide data to define if the combination is effective and when they should be used.

Rather that use CAR-T in combination with check-point inhibitors, strategies are focused to genetically engineer disruption of the PD-1 pathway into CAR-T themselves (Ankri et al., 2013; Kobold et al., 2015; Prosser et al., 2012).

8. Regulatory improvements

Although the regulatory aspects define objective evidence and safety rules for providing the most strong and relevant results, the general rules often limit the development of the improvements, blocking some suggestive implementations that could be useful for patients. If we want to introduce and improve CAR therapy, these rules, mainly defined for the development of conventional drugs, should be adapted to the real needs and peculiarities of this cell immunotherapy. In fact, CAR-T products are autologous cells modified by a procedure, being more relevant for the success of the treatment the initial state of the immune system of the patient, than the own characteristics (doses, cell composition, in vitro function, ...) of the "product"; some patients can obtain complete responses with low number of transduced cells, being the most promising data about cell-subpopulations just statistical positive tendencies no-predictive for a specific patient. To be strict and too detailed in this characterization of the final CAR-T ATMP, can avoid the development and use of some products that can be useful for the specific patient, and some inflexible rules (such the definition of a single dose instead of the use of fractionated doses in some patients) can produce unnecessary adverse effects in the patients such have been demonstrated in trials using fractionated doses (Ortíz-Maldonado et al., 2021).

9. Pharma and Academic collaboration for supplying sustainable options to the health systems

Finally, although almost all the initial proposals of CAR therapy were initially developed by academic developments and later were transferred to commercial pharmaceutical companies, in the last years the power of these companies are conditioning the development of different academic proposals. Pharma industry can allow that the effective CAR products arrive around the world to almost all the patients, but the prices of the current CAR-T therapy make these commercial products unattainable for many patients and health systems with limited resources. Especially relevant is the development of treatment for infrequent diseases, where the reimbursements for the high inversion needed for industry will block the development of these unprofitable proposals. Just if academic proposals are accepted, these treatments will arrive to patients and pharmaceutical companies should accept and support this fair competition. This support to academic development and use in some diseases and indications would be especially relevant if the industry understand that regulatory rules should adapt to help in the development of academic proposals. In Europe, hospital exemption development (Trias et al., 2022) could be one of the ways to make effective these non-profitable treatments. If industry do not block these proposals, possibly the most profitable products could be transferred to pharmaceutical companies in a win-to-win way.

10. Final discussion and conclusion

Although most of the work in CAR immunotherapy is focused on the use of a conventional structure (scFv next to two o three signaling domains), autologous cytotoxic cells (mainly T and NK cells), and infusion in a single dose under rules defined for commercial products, new options can change this usual model. A wide range of engineering strategies are being addressed to improve the safety, efficacy, and applicability of this treatment. Radical changes are complex but probably they are not only possible but also necessary to allow the growth of a disruptive field that could provide very personalized treatments for solving most of the medical challenges associated to cancer and why not to a high number of immune-mediated diseases. Try to restrict a new therapy to "initial and old" ways of development under "too limiting rules" can reduce its high potentiality for solving medical problems. New proposals, methods, approaches and rules for CAR-related treatments can provide the way to convert a challenge in the greatest opportunity for improving the health of cancer and other non-cancer patients. In our healthcare environment, it is important to have in mind that most of our health systems needs new options to make possible the access of population to the most effective treatments and commercial proposals (under a absolutely free market) are of complex fit within our health systems in the welfare state that is frequently a sign of identity of Europe.

Acknowledgments

Work developed thanks to grants PI18/000775, AC18/00072 (CE_ERA-Net_Nanomed18) and RD21/0017/0019 (RICORS-TERAV) from *Instituto de Salud Carlos III* with *Fondo Europeo de Desarrollo Regional (FEDER) "una manera de hacer Europa"* and by the "Fundació bancària la Caixa" (CP042702/LCF/PR/GN18/50310007).

References

Ankri, C., Shamalov, K., et al., 2013. Human T cells engineered to express a programmed death 1/28 costimulatory retargeting molecule display enhanced antitumor activity. J. Immunol. 191 (8), 4121–4129.

Brentjens, R.J., Rivière, I., et al., 2011. Safety and persistence of adoptively transferred autologous CD19-targeted T cells in patients with relapsed or chemotherapy refractory B-cell leukemias. Blood 118 (18), 4817–4828.

Burki, T.K., 2021. CAR T-cell therapy roll-out in low-income and middle-income countries. Lancet Haematol 8 (4), e252–e253.

Chmielewski, M., Hombach, A.A., Abken, H., 2014. Of CARs and TRUCKs: chimeric antigen receptor (CAR) T cells engineered with an inducible cytokine to modulate the tumor stroma. Immunol. Rev. 257 (1), 83–90.

Chong, E.A., Melenhorst, J., et al., 2017. Phase I/II study of pembrolizumab for progressive diffuse large B cell lymphoma after anti-CD19 directed chimeric antigen receptor modified T cell therapy. Blood 130 (Suppl. 1), 4121.

Couzin-Frankel, J., 2013. Breakthrough of the year 2013. Cancer immunotherapy. Science 342 (6165), 1432–1433.

Dwivedi, A., Karulkar, A., et al., 2019. Lymphocytes in cellular therapy: functional regulation of CAR T cells. Front. Immunol. 9, 3180.

Eyquem, J., Masilla-Soto, J., et al., 2017. Targeting a CAR to the TRAC locus with CRISPR/Cas9 enhances tumour rejection. Nature 543 (7643), 113–117.

Good, C.R., Aznar, M.A., et al., 2021. An NK-like CAR T cell transition in CAR T cell dysfunction. Cell 184 (25). 6081–6100.e26.

Han, L., Zhang, J.-S., et al., 2021. Single VHH-directed BCMA CAR-T cells cause remission of relapsed/ refractory multiple myeloma. Leukemia 35 (10), 3002–3006.

Haque, S., Vaiselbuh, S.R., 2021. CD19 chimeric antigen receptor-exosome targets CD19 positive B-lineage acute lymphocytic leukemia and induces cytotoxicity. Cancer 13 (6), 1401.

Hombach, A.A., Rappl, G., Abken, H., 2013. Arming cytokine-induced killer cells with chimeric antigen receptors: CD28 outperforms combined CD28-OX40 "superstimulation". Mol. Ther. 21 (12), 2268–2277.

Jensen, M.C., Riddell, S.R., 2015. Designing chimeric antigen receptors to effectively and safely target tumors. Curr. Opin. Immunol. 33, 9–15.

Kawalekar, O.U., O'Connor, R.S., et al., 2016. Distinct signaling of coreceptors regulates specific metabolism pathways and impacts memory development in CAR T cells. Immunity 44 (2), 380–390.

Klichinsky, M., Ruella, M., et al., 2020. Human chimeric antigen receptor macrophages for cancer immunotherapy. Nat. Biotechnol. 38 (8), 947–953.

Kobold, S., Grassmann, S., et al., 2015. Impact of a new fusion receptor on PD-1-mediated immunosuppression in adoptive T cell therapy. J Natl Cancer Inst 107 (8), djv146 (2015).

Kochenderfer, J.N., Wilson, W.H., et al., 2010. Eradication of B-lineage cells and regression of lymphoma in a patient treated with autologous T cells genetically engineered to recognize CD19. Blood 116 (20), 4099–4102.

Li, A.M., Hucks, G.E., et al., 2018. Checkpoint inhibitors augment CD19-directed chimeric antigen receptor (CAR) T cell therapy in relapsed B- cell acute lymphoblastic leukemia. Blood 132 (Suppl. 1), 556.

Liu, E., Marin, D., Banerjee, P., et al., 2020. Use of CAR-transduced natural killer cells in CD19-positive lymphoid tumors. N. Engl. J. Med. 382 (6), 545–553.

Lynn, L.C., Weber, E.W., et al., 2019. C-Jun overexpression in CAR T cells induces exhaustion resistance. Nature 576 (7786), 293–300.

Magnani, C.F., Tettamanti, S., et al., 2020. Transposon-based CAR T cells in acute leukemias: where are we going? Cell 9 (6), 1337.

Muliaditan, T., Halim, L., et al., 2021. Synergistic T cell signaling by 41BB and CD28 is optimally achieved by membrane proximal positioning within parallel chimeric antigen receptors. Cell Rep Med 2 (12), 100457.

Ortíz-Maldonado, V., Rives, S., et al., 2021. CART19-BE-01: a multicenter trial of ARI-0001 cell therapy in patients with CD19+ relapsed/refractory malignancies. Mol. Ther. 29 (2), 636–644.

Pegram, H.J., Lee, J.C., et al., 2012. Tumor-targeted T cells modified to secrete IL-12 eradicate systemic tumors without need for prior conditioning. Blood 119 (18), 4133–4141.

Porter, D.L., Levine, B.L., et al., 2011. Chimeric antigen receptor-modified T cells in chronic lymphoid leukemia. N. Engl. J. Med. 365 (8), 725–733.

Prosser, M.E., Brown, C.E., et al., 2012. Tumor PD-L1 co-stimulates primary human CD8+ cytotoxic T cells modified to express a PD1:CD28 chimeric receptor. Mol. Immunol. 51 (3–4), 263–272.

Rafiq, S., Hackett, C.S., Brentjens, R.J., 2020. Engineering strategies to overcome the current roadbloks in CAR T cell therapy. Nat. Rev. Clin. Oncol. 17 (3), 147–167.

Rotolo, R., Leuci, V., et al., 2019. CAR- based strategies beyond T lymphocytes: integrative opportunities for cancer adoptive immunotherapy. Int. J. Mol. Sci. 20 (11), 2839.

Sebestyen, Z., Prinz, I., Déchanet-Merville, J., Silva-Santos, B., Kuball, J., 2020. Translating gammadelta (gammadelta) T cells and their receptors into cancer cell therapies. Nat. Rev. Drug Discov. 19 (3), 169–184.

Trias, E., Juan, M., Urbano-Ispizua, A., Calvo, G., 2022. The hospital exemption pathway for the approval of advanced therapy medicinal products: an underused opportunity? The case of the CAR-T ARI-0001. Bone Marrow Transplant. 57 (2), 156–159.

Tristán-Manzano, M., Maldonado-Pérez, N., et al., 2021. Physiological (TCR-like) regulated lentiviral vectors for the generation of improved CAR-T cells. medRxiv. https://doi.org/10.1101/2021.03.17.21253300.

Weber, E.W., Parker, K.R., et al., 2021. Transient rest restores functionality in exhausted CAR-T cells through epigenetic remodeling. Science 372 (6537), 49.

CHAPTER SIX

Adoptive tumor infiltrating lymphocyte transfer as personalized immunotherapy

Ines Diaz-Cano[a], Luis Paz-Ares[a,b,c], and Itziar Otano[a,b,*]

[a]H12O-CNIO Lung Cancer Clinical Research Unit, Health Research Institute Hospital 12 de Octubre/Spanish National Cancer Research Center (CNIO), Madrid, Spain
[b]Spanish Center for Biomedical Research Network in Oncology (CIBERONC), Madrid, Spain
[c]Medicine and Physiology Department, School of Medicine, Complutense University of Madrid, Madrid, Spain
*Corresponding author: e-mail address: iotanoan@alumni.unav.es

Contents

1. Introduction	164
2. History of TILs	165
3. Clinical studies	166
3.1 Evidences of lymphodepletion regimens	169
4. TIL manufacture	170
5. Antigen-specificity of TILs	172
5.1 Neoantigens	174
5.2 TIL selection based on cell phenotype	176
6. T-cell differentiation state of TILs	178
7. Combinatorial therapies	181
8. Synthetic TILs in cancer therapy	182
9. Perspectives and conclusions	184
Acknowledgments	185
Author contributions	185
Competing interests	185
References	185

Abstract

Cancer is a leading cause of death worldwide and, despite new targeted therapies and immunotherapies, a large group of patients fail to respond to therapy or progress after initial response, which brings the need for additional treatment options. Manipulating the immune system using a variety of approaches has been explored for the past years with successful results. Sustained progress has been made to understand the T cell-mediated anti-tumor responses counteracting the tumorigenesis process. The T-lymphocyte pool, especially its capacity for antigen-directed cytotoxicity, has become a central focus for engaging the immune system in defeating cancer. The adoptive cell transfer of autologous tumor-infiltrating lymphocytes has been used in humans for over

30 years to treat metastatic melanoma. In this review, we provide a brief history of ACT-TIL and discuss the current state of ACT-TIL clinical development in solid tumors. We also discuss how key advances in understanding genetic intratumor heterogeneity, to accurately identify neoantigens, and new strategies designed to overcome T-cell exhaustion and tumor immunosuppression have improved the efficacy of the TIL-therapy infusion. Characteristics of the TIL products will be discussed, as well as new strategies, including the selective expansion of specific fractions from the cell product or the genetic manipulation of T cells for improving the in-vivo survival and functionality.

In summary, this review outlines the potential of ACT-TIL as a personalized approach for epithelial tumors and continued discoveries are making it increasingly more effective against other types of cancers.

Abbreviations

ACT	adoptive T-cell
APCs	antigen-presenting cells
CR	complete response
CRISPR	clustered regularly interspaced short palindromic repeats
CTLA-4	cytotoxic T-lymphocyte antigen 4
DNA	deoxyribonucleic acid
ELISA	enzyme-linked immunosorbent assay
HNSCC	head and neck squamous cell carcinoma
ICD	immunogenic cell death
IL-2	interleukin-2
IL-7	interleukin-7
IL-12	interleukin-12
IL-15	interleukin-15
MDSCs	myeloid suppressive cells
NSCLC	non-small cell lung cancer
OR	overall response
PD-1	programmed death-1
RNA	ribonucleic acid
TBI	total body irradiation
TILs	tumor-infiltrating lymphocytes
TLR4	toll like receptor 4
TME	tumor microenvironment
Tregs	regulatory T-cell

1. Introduction

An abundance of intratumoral lymphocytes is typically a positive prognostic indicator (Linette and Carreno, 2019), suggesting that tumor-infiltrating immune cells play an important role in tumor eradication. TILs are a group of lymphocytes that have naturally penetrated the tumor microenvironment (TME) and remain actively fighting the tumor. Many of these are T cells capable of recognizing tumor-specific antigens and neoantigens. Adoptive T-cell (ATC) therapy based on the infusion of autologous tumor-infiltrating lymphocytes (TILs) into patients with cancer has shown considerable promise in the past years. The viability of this type of therapy was first shown in 1980, where tumor regression was demonstrated in selected patients receiving adoptive transfer of lymphokine-activated killer cells in combination with recombinant IL-2 (Rosenberg et al., 1988). Significant responses can be achieved with the therapeutic use of these cells, predominantly in patients with metastatic melanoma. However, the expansion and re-infusion of these cells is a promising approach.

Here, we aim to examine and discuss the history, the current state of ACT for the treatment of human cancer, as well as the principles of effective treatment that point toward improvements in this approach.

2. History of TILs

Allogenic hematopoietic stem cell transplants for leukemia represented the first effective adoptive transfer approach deployed clinically, and clinical improvement was shown to be mediated by a T-cell graft versus tumor response (Weiden et al., 1979). The importance of T cells in cancer progression was firstly observed following adoptive transfer of syngeneic lymphocytes from rodents heavily immunized against the tumor (Delorme and Alexander, 1964; Fefer, 1969; Gross, 1943), in which a modest growth inhibition of small established tumors was observed.

In vitro studies in 1986 showed that human TILs obtained from resected melanomas contained cells capable of specific recognition of autologous tumor (Muul et al., 1987). Nowadays we know that tumor-specific T cells are activated through exposure to tumor-associated antigens presented by professional antigen-presenting cells (APCs), and these activated T cells are capable of directly recognizing antigens loaded on the surface of tumor cells. Thus, lymphocytes infiltrating into the stroma of tumors represent a concentrated source of lymphocytes capable of recognizing tumor cells.

The discovery of interleukin-2 (IL-2) as an imperative agent in the function and response of immune cells allowed for the long-term culture, maintenance and expansion of T cells. Studies in murine tumor models and pilot studies in patients with advanced cancer demonstrated that the adoptive transfer of these syngeneic tumor-infiltrating lymphocytes (TILs) expanded in vitro in combination with IL-2 could mediate regression of established tumors (Kradin et al., 1987; Rosenberg et al., 1986; Topalian et al., 1988).

3. Clinical studies

Nowadays, the vast majority of clinical trials that are being developed are focused on metastatic melanoma (Table 1). Even though, ACT-TIL has not been FDA approved yet for solid malignancies. Based on combined data from seven individual studies with 332 patients in total, the average objective response rate for ACT-TIL and high IL-2 therapy was 44% and durable complete responses were reported in 49 patients.

A number of clinical trials have reported encouraging results (Table 1). Lifileucel (LN-145, LN-144 and LN-145-S1), developed by Iovance Biotherapeutics is an autologous, centrally manufactured tumor-infiltrating lymphocyte product for the treatment of metastatic or unresectable melanoma, squamous cell carcinoma, recurrent or metastatic non-small cell lung cancer, recurrent or metastatic head and neck squamous cell carcinoma (HNSCC) or refractory/relapsed non-small cell lung cancer (NSCLC) (Table 1). A prospective phase 2 study designed to evaluate efficacy and safety of TIL (LN-144) on metastatic melanoma patients was completed. LN-144 demonstrated an ORR of 36% and disease control rate of 80% with a consistent safety profile (Sarnaik et al., 2020). More recently, the LN-145, also developed by Iovance Biotherapeutics demonstrated 44% ORR in a prospective phase 2 study that included patients with recurrent, metastatic or persistent cervical carcinoma, which received TIL (LN-145) based ACT, IL-2 and lymphodepletion regimens (Jazaeri et al., 2019).

Current ACT trials have coupled TIL infusions with efforts to manipulate the host microenvironment through preparative and post-infusion regimens as to promote TIL activation and proliferation after transfer. Specifically, pre-infusion lymphodepletion with chemotherapy (fludarabine or cyclophosphamide) or total body irradiation (TBI) previous to TIL infusion led to objective response rates ranging from 38% to 50% in patients with metastatic melanoma. Durable and complete tumor regression was observed in up to 22% of patients (Besser et al., 2013; Goff et al., 2016; Pilon-Thomas et al., 2012; Radvanyi et al., 2012; Rosenberg et al., 2011; Topalian et al., 1988) and 20% have not experienced recurrences at follow-up times of 5–10 years. Increasing lymphodepletion by TBI prior to TIL infusion showed no additional benefit as it was shown in a randomized study of patients with metastatic melanoma treated with or without addition of TBI with

Table 1 Selected clinical trials of TILs-ACT.

	Cancer histology	Year	Patients	Number of ORs	Notes	References
	Melanoma	1988	20	55%	Original use TIL ACT	Rosenberg et al. (1988)
	Metastatic ovarian cancer	1991	13		3-year disease-free survival rate of 82.1%	Fujita et al. (1995)
	Melanoma	1994	86	34%		Rosenberg et al. (1994)
	Melanoma	2002	13	46%	Lymphodepletion before cell transfer	Dudley et al. (2002b)
	Melanoma	2011	93	56%	20% CR beyond 5 years	Rosenberg et al. (2011)
	Metastatic melanoma	2010	25	42%	Low dose of IL-2	Andersen et al. (2016)
	Melanoma	2012	31	48%		Radvanyi et al. (2012)
	Metastatic melanoma	2012	13	26%	Intention to treat: 26% OR rate	Pilon–Thomas et al. (2012)
	Melanoma	2013	54	40%		Besser et al. (2013)
	Cervical cancer	2014	9	33%		Tran et al., (2014)
	Bile duct	2014	1		Mutated ERB2	Tran et al. (2014)
	Metastatic melanoma	2015	178	36%	LN-144	Sarnaik et al. (2020)
	Metastatic melanoma	2016	51	54%		Goff et al. (2016)
	Cervical cancer	2017	27	44%	LN-145	Jazaeri et al. (2019)
In vitro sensitization	Melanoma	2008	9	33%	NY-ESO-1	
	Leukemia	2014	11		WT1	Chapuis et al. (2013)
	Epithelial tumors	2021			Enriched for neoantigens	

Continued

Table 1 Selected clinical trials of TILs-ACT.—cont'd

	Cancer histology	Year	Patients	Number of ORs	Notes	References
Genetic modifications	Metastatic melanoma	2010	34	63%	IL-12	Zhang et al. (2015)
	Metastatic melanoma	2003	33		IL-2	
	Metastatic melanoma	2014	6	16%	4-1BB selected TILs	
Combination therapies	Metastatic melanoma	2012	13	38	Ipilimumab	Mullinax et al. (2018)
	Metastatic melanoma Metastatic ovarian cancer	2013			Ipilimumab	
	Ovarian cancer	2017	7	33%	Nivolumab	Kverneland et al. (2020)
	NSCLC	2017	20		Nivolumab	
	Metastatic ovarian cancer	2017			Ipilimumab + Nivolumab	
	Cervical carcinoma Advanced ovarian cancer Malignant melanoma	2017			Pembrolizumab	
	Metastatic melanoma SCCHN NSCLC	2018			Pembro/Ipilimumab + Nivolumab	
	Metastatic melanoma	2018			Nivolumab + IFNα	
	Metastatic melanoma	2018			Nivolumab	
	Metastatic ovarian cancer	2020			Nivolumab + Relatlimab	

lymphodepleting chemotherapy. Both groups of patients presented similar complete response (CR) rates and overall response (OR) (Goff et al., 2016).

3.1 Evidences of lymphodepletion regimens

The benefits of lymphodepletion are in part by increasing the availability of homeostatic cytokines, such as IL-7 and IL-15 (Wrzesinski et al., 2010). These cytokines are produced by non-lymphoid sources in response to lymphopenia, where IL-7 is required for the proliferation and survival of T cells and IL-15 maintains and improves the proliferation of T cells (Schluns et al., 2000, 2002), particularly of memory T cells. The competition for cytokines between host and transferred T cells is known as the "cytokine sink" effect. The transient lymphodepletion of endogenous lymphocytes and NK cells leads to an increase in serum levels of IL-7 and IL-15 that has been linked to long-term persistence of transferred TILs and better outcomes (Chapuis et al., 2012; Gattinoni et al., 2005a; Wrzesinski et al., 2010).

The depletion of host lymphocyte-mediated immunosuppressive cells that potentially hinder the functions of transferred T cells has also been attributed to lymphodepletion regimens. Lymphodepletion with chemotherapy and/or radiation preparative regimens have shown to be effective in eliminating myeloid suppressive cells (MDSCs) and regulatory T-cell (Tregs) populations in tumor-bearing mice; and enhancing anti-tumor immune activity (Vincent et al., 2010; Zhang et al., 2005). However, radiotherapy (RT) favors the infiltration by Treg and myeloid cells through the release of radiation-induced chemokines (CSF-1, CCL2, CXCL12) (Kozin et al., 2010). Although recruitment of immunosuppressive cells may be transient and be later replaced by an influx of effector T cells (Filatenkov et al., 2015).

However, these mechanisms might not fully account for the dramatically improved tumor regression resulting from lymphodepleting preparative. Interestingly, innate immune activation via microbial-derived agonists, liberated from radiation, was responsible for triggering toll like receptor 4 (TLR4) signaling and enhancing ACT effectiveness in mice (Nelson et al., 2016). Moreover, radiotherapy triggers immunogenic cell death (ICD), a type of cell death characterized by the release of danger signals, which elicit an effective costimulation and presentation of tumor antigens and priming of antigen-specific T cells (Zhang et al., 2007).

4. TIL manufacture

TIL therapy begins with full or partial surgical resection of the tumor. Accessibility is not frequently a hurdle in case of metastatic melanoma as these tumors often metastasize to lymph nodes, cutaneous, and subcutaneous sites. Tumors from other cancer types, however, can be difficult to access surgically, or are complicated to sterilely obtain. For example, gastrointestinal cancers are specifically challenging to resect without bacteria contamination.

The ex vivo expansion phases required to produce large quantities of phenotypically effective TILs presents a great challenge. The standard method for the expansion of TILs is organized in two phases, the first procedure consists in a pre-Rapid Expansion Protocol (preREP) phase, in which tumors are surgically excised and cut in small fragments or enzymatically digested in order to achieve cell suspensions (Kumar et al., 2021). Then, high dosage (6000 IU/mL) of IL-2 has to be added to every tumor fragment in a cell culture plate (Wang and Rivière, 2015). After 2–3 weeks, lymphocytes are tested for reactivity against tumors, if available, by measuring interferon-gamma secretion by enzyme-linked immunosorbent assay (ELISA) (Rosenberg and Restifo, 2015). The solely use of IL-2 for cell expansion resulted in a limited expansion of TILs, and due to the DNA mapping of the CD3 complex, an anti-CD3 (OKT3) agonist antibody was designed for stimulating the TCR/CD3 complex. Reactive TILs are then pooled and expanded in the presence of IL-2 (3000 IU/mL), anti-CD3 (OKT3) antibody and irradiated allogeneic peripheral blood mononuclear cells (PBMCs), feeder cells, for 14 days in the REP phase. TILs are then harvested, concentrated, cryopreserved or infused intravenously into the patient (Wang and Rivière, 2015). A typical REP results in 1000- to 2000-fold expansion of TILs during the 14-days culture period. In summary, around 5–6 weeks from surgical procedure are needed to obtain the minimal cell value (up to 10^{11}) for being infused into the patient (Rosenberg and Restifo, 2015) (Fig. 1).

In an attempt to shorten pre-REP culture time and increase cell numbers, manipulation of co-stimulatory signaling pathways was examined. Preclinical studies demonstrated that TILs isolated from solid tumors express the co-stimulatory molecule CD137, which can be exploited ex vivo to enhance TILs expansion through direct addition of anti-CD137 agonist antibodies to tumor fragments in culture (Chacon et al., 2015a). The addition of agonist

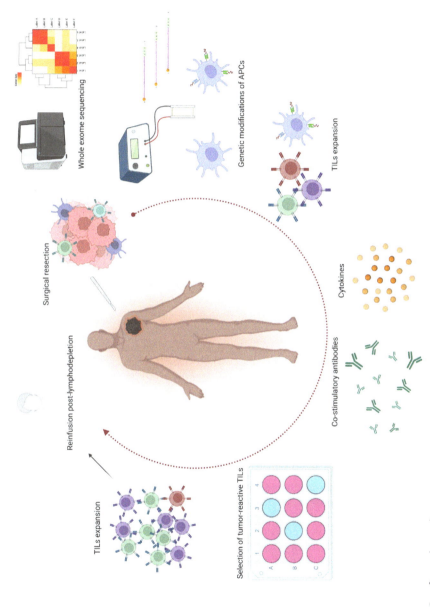

Fig. 1 See figure legend on next page.

anti-CD137 antibody during pre-REP resulted in increased survival and expansion of effector TILs cultured *ex vivo* (Chacon et al., 2015b) (Fig. 1).

A positive association with longer culture times and shorter telomere length of TILs was found (Zhou et al., 2005a). To overcome the excessive differentiation of the TIL product after 6-weeks culture time, some studies have simplified the process to culture "young" TILs (<20 days culture time). "Young" TIL cultures are made of bulk lymphocytes rather than individual microcultures, and the tumor recognition screening assay is eliminated from the process (Donia et al., 2012; Dudley et al., 2010). Interestingly, clinical response rates with "young" TILs are comparable with larger-culture time TILs, which makes "young" TILs the current standard in the field (Dudley et al., 2010; Itzhaki et al., 2011).

Significant improvements made to the manufacturing process have involved a reduction on the cycle time by shortening the culture period but without adversely affecting product quality or function. Ongoing efforts to improve the TIL manufacturing process will maximize the high-quality of the cell therapy product at large-scale.

5. Antigen-specificity of TILs

TILs have shown impressive results in patients with metastatic melanoma, however, the response rates in other epithelial cancers such as breast (Lee et al., 2017), ovarian (Freedman et al., 1994), renal (Andersen et al., 2018; Kradin et al., 1989), liver (Jiang et al., 2015) or cervical (Stevanovic et al., 2015, 2019) tumors have been very modest. Clinical benefit is hampered by a lack of T-cell persistence in vivo, poor T-cell functionality and insufficient localization of infused lymphocytes at the tumor site.

TIL products are generally mixtures of highly polyclonal CD8+ and CD4+ T cells, and it is now well understood that CD4+ T cells are crucial

Fig. 1 Autologous tumor-infiltrating lymphocytes for adoptive cell transfer in cancer patients. The resected specimen is divided into multiple tumor fragments that are individually grown in IL-2. For the "non-specific" TIL therapy the individual cultures are then moved to a rapid expansion protocol before reinfusion into patients. Neoantigen-TIL therapy involves the sequencing of exomic or whole-genome DNA from tumor cells and healthy cells to identify tumor-specific mutations. Corresponding minigenes or peptides encoding each mutated amino acid are synthesized and expressed in or pulsed into a patient's autologous antigen-presenting cells (APCs) for presentation in the context of a patient's HLA. Individual mutations responsible for tumor recognition are identified by analyzing activation of a T-cell co-stimulatory marker, such as CD137 or IFNγ production in co-cultures assays. Then, a large-scale of selected TILs is produced.

for assisting CD8+ TILs (Ahrends et al., 2017, 2019; Borst et al., 2018). A TIL-based ACT benefits from a multitarget T-cell attack directed against multiple, different and largely unknown antigens. The intrinsic capacity of intratumoral T cells to recognize autologous tumor antigens can be rare and variable (~10% in the case of melanoma, ovarian and colorectal cancer TILs); whereas the majority of TILs are bystander T cells (Andersen et al., 2012; Scheper et al., 2019; Simoni et al., 2018). The capacity of bystander TILs to respond to inflammation could have important implications for cancer therapy aiming to reinvigorate tumor-specific responses (Danahy et al., 2020; Meier et al., 2022). However, the relevance of bystander cells to anti-cancer immunity during tumorigenesis as well as in later stages of cancer and during immunotherapy remains unclear. Thus, strategies to enrich cultures for tumor-antigen specificities may increase the chances of obtaining a final TIL product with optimal tumor reactivity as well as persistence in vivo.

Among tumor associated antigens (TAA), three major groups of antigen specificity have been described: cancer germline antigens and differentiation antigens (which can also be present in normal tissues but in a restricted distribution) and oncogenic viral antigens (that arise from viral proteins in viral-associated cancers). Tissue differentiation antigens are encoded by genes that are expressed by specific tumors and can be present in normal tissue but in a restricted distribution. Examples of this class of TAA include mammaglobin-A, which is highly expressed in breast cancers; prostate-specific antigen (PSA), which is expressed in the prostate gland and prostate cancer; melanoma antigen 1 (MART1), melanocyte protein PMEL and tyrosinase, which are expressed by normal melanocytes and melanoma cells (Buonaguro et al., 2011). However, as these antigens are also expressed in healthy tissue, natural TCR binding is often of low affinity as a result of negative selection of high-affinity T cells in the thymus. Cancer–testis antigens are a subset of antigens that are not expressed in normal adult tissues, except by germline and trophoblastic cells, but are highly expressed across cancers. The best studied antigens are the melanoma-associated antigen (MAGE) family, sarcoma antigen1 (SAGE1), cancer–testis antigen 1 (NY-ESO-1) and the oncofetal antigen 5T4 (Chen et al., 1997; Coulie et al., 2014; Stern and Harrop, 2017). Oncogenic viral antigens have been identified in virus-induced cancers such as human papillomavirus (HPV)-associated cervical cancer, hepatitis B virus-associated hepatocellular carcinoma, Merkel cell polyomavirus (MCPyV) and human herpesvirus 8-associated Kaposi sarcoma (Anders et al., 2011; Feng et al., 2008; Kenter et al., 2009; Saxena et al., 2021).

The identification of epitopes in tumor-associated antigens led to a specific expansion of T-cell clones. The infusion of the TIL product resulted in clinical benefit and an in vivo long-term persistence that may be a predictor of response (Chapuis et al., 2012, 2014; Heslop et al., 2010; Hunder et al., 2008; Stevanovic et al., 2019, 2015; Tran et al., 2016). All of the above-described TAAs are subjected to some degree of central tolerance and lack complete specificity to the tumor.

5.1 Neoantigens

Nowadays, several studies demonstrated that the effectiveness of TIL-based ACT in metastatic melanoma is based on the specific recognition of neoantigens (Prickett et al., 2016; Robbins et al., 2013). Direct evidence of the therapeutic effect of targeting neoantigens came from pre-clinical studies in which vaccination with neoantigens led to tumor reduction in mouse models (Castle et al., 2012; Gubin et al., 2014; Kreiter et al., 2015; Matsushita et al., 2012; Yadav et al., 2014). Emerging studies suggested that higher neoantigen tumor load is associated with stronger T-cell responses and better clinical outcome in patients treated with immune checkpoint blockade (Hugo et al., 2016; Snyder et al., 2014; van Allen et al., 2015).

Each patient's cancer is driven by a unique set of mutations (the mutanome) that arise during the progression toward a malignant cell (Ciriello et al., 2013; Lawrence et al., 2013). Neoantigens are derived from a somatic gene mutation and gene fragment insertion or deletion that result in the formation of novel protein sequences that can be displayed by HLA molecules. As compared with nonmutated self-antigens, neoantigens have been postulated to be of particular relevance to tumor control, as the quality of the T-cell pool that is available for these antigens is not affected by central T-cell tolerance. Mutant tumor-antigen-specific T cells are present in progressively growing tumors, and they are reactivated following treatment with anti-PD-1 and/or anti-CTLA-4 blocking antibodies rendering them capable of mediating tumor rejection (Gubin et al., 2014, 2015; Le et al., 2017). Since tumor mutational burden (TMB) is not perfectly predictive, more is to be learned about how to efficiently identify neoantigen-associated profiles. However, despite ~90% of cancer incidences globally being caused by solid tumors, these studies have rarely been extended to non-melanoma cancers, due to a lower immunogenicity of these tumors. More recently, a big effort is being made to identify shared immunogenic, poly-epitope frameshift-peptides likely to be common among multiple patients (Roudko et al., 2020).

Many efforts are underway to identify and expand neoantigen-reactive T cells for personalized therapies. Preliminary findings have suggested that around ~1–2% of tumor mutations are transduced into epitopes that bind to available MHC molecules on tumor cells and can be recognized by the T-cell repertoire, indicating that a low number of mutations may reduce the chances of finding functional neoantigen-specific T cells (Rosenberg and Restifo, 2015).

Neoantigen-reactive TCRs that can be harnessed to specifically target tumor cells, avoiding off-target cytolysis of healthy tissue have been identified (Paria et al., 2021). Recently, the Tumor Neoantigen Selection Alliance (TESLA), a scientist consortium, pointed out the urgent need for a harmonized approach to seek tumor epitope immunogenicity, neoantigen prediction, to ultimately generate a broadly accessible reference dataset for personalized cancer treatments (Wells et al., 2020).

Whole exome sequencing can be used to analyze individual tumor-normal pairs in patients to consistently identify nonsynonymous cancer mutations (Fig. 1). For all mutations that resulted in the formation of novel protein sequences, potential MHC binding peptides were predicted by algorithms, and the resulting set of potential neoantigens was used to evaluate T-cell reactivity (Castle et al., 2012; Matsushita et al., 2012; Wells et al., 2020). However, among the large number of predicted neoantigens, only a minority is recognized by autologous patient T cells, and strategies to broaden neoantigen-specific T-cell responses are therefore attractive.

Another method to identify potential neoantigens is based in Tandem minigenes (TMG) libraries encoding point mutated amino acids flanked by 10–12 amino acids. Sequences of 6–20 minigenes are then linked into tandem minigenes, and these DNA constructs are subsequently cloned into an expression plasmid and in vitro transcribed to RNA, which is electroporated into the patient's autologous APCs. Culture of the patient's TILs with these APCs can identify the tandem minigene as well as the individual minigene responsible for tumor recognition (Lu et al., 2014).

Recent advances in T-cell technologies have led to the possibility of an extensive screening of T-cell recognition against large libraries of patient-derived neoepitopes (Gartner et al., 2021). The use of DNA-barcoded multimers could also be applied to identify and expand neoantigen-reactive TILs (Kristensen et al., 2022). The identification of antigen-specific T cells based on an interaction-dependent fucosyl-biotinylation has been proposed as alternative method to bioinformatics-assisted processes (Liu et al., 2020).

Adoptive transfer of autologous T-cell products containing high fractions of neoantigen-specific T cells has generated tumor regression in a range of cancer types (Tran et al., 2014; Zacharakis et al., 2018) and has further shown the ability to mediate durable complete regression in a substantial fraction of patients with metastatic melanoma (Dudley et al., 2008; Huang et al., 2004; Lu et al., 2013; Robbins et al., 2013; Tran et al., 2014, 2016; Zhou et al., 2005b). Based on these observations, the clinical efficacy of the ACT-TIL therapy is strongly dependent on the quality of T cells to recognize and eradicate the tumor cells. The combination of high-throughput genomics and cellular immunotherapy platforms permits the identification of somatic alterations and the prediction of potential neoantigens that could be utilized as targets for therapy by vaccination or adoptive TIL transfer.

5.2 TIL selection based on cell phenotype

The identification of appropriate tumor antigens is one of the greatest challenges in the field of cancer immunotherapy. The identification of surrogate markers that exclude bystander T cells from antigen-specific T cells will narrow the heterogeneous population of T cells to a highly specific TIL repertoire. In addition, the subsequent characterization of the different functional, phenotypical and transcriptional signature profiles of the tumor-reactive TIL pool is an important issue for improving and developing new immunotherapies. Following antigen stimulation, T cells undergo dynamic functional and phenotypic changes, and upregulate surface expression of multiple activation-associated molecules.

In melanoma, PD-1 was shown to be a biomarker of functional tumor-antigen reactive T lymphocytes (Gros et al., 2014). Even though high levels of PD-1 are related to T-cell exhaustion, PD-1 expression on TILs also accurately identify the repertoire of clonally expanded tumor-reactive cells compared to the corresponding bulk or PD-1 negative fractions (Gros et al., 2014; Salas-Benito et al., 2021). The infusion of selected PD-1 positive T cells showed a robust anti-tumor response in several mice models (Fernandez-Poma et al., 2017; Jing et al., 2017).

CD137-expressing TILs were shown to be enriched for antigen-specific tumor T cells in melanoma (Ye et al., 2014). CD137 (4-1BB, TNFSFR9) is a TNFR-family member that promotes T-cell proliferation and survival functions on T cells and was originally identified as an inducible molecule expressed on activated mouse and human T cells (Shuford et al., 1997;

Watts, 2005). A superior anti-tumor response was observed after the infusion of CD137^{pos} TILs compared to CD137^{neg} TILs in a melanoma mouse model (Parkhurst et al., 2017).

More recently, it has been proposed that CD39 could be useful as a marker of tumor-specific CD8 T cells (Krishna et al., 2020; Simoni et al., 2018). CD39 is an ectonucleotidase expressed on activated T cells (Kansas et al., 1991), which, in coordination with CD73 can result in local production of adenosine leading to an immunosuppressive environment (Moesta et al., 2020). CD39 was identified as a marker for exhausted T cells in patients with chronic viral infections (Gupta et al., 2015), is upregulated on activated CD8 T cells after chronic antigenic stimulation at the tumor site and correlates with clinical beneficial parameters in colorectal cancer, lung cancer, and head and neck cancer (Duhen et al., 2018; Simoni et al., 2018; van den Bulk et al., 2019). Moreover, the co-expression of CD39 and CD103 identifies a unique population of TILs restricted to the tumor microenvironment (Duhen et al., 2018). This cell subset shows a resident memory (T_{RM}) phenotype, expresses high levels of exhaustion markers, but is enriched of tumor-reactive cells recognizing and killing autologous tumor cells (Kortekaas et al., 2020). Extensive studies have shown that even though the PD-1^{high} subpopulation of CD39$^+$ TILs showed features of terminally exhausted T cells, a significant portion expressed CD137, suggesting that they may retain higher proliferative and reinvigoration potential, providing anti-tumor responses (Leem et al., 2020).

However, this definition does not exclude that a fraction of the tumor-specific T cells could also be found in the CD39$^-$ pool. In one study, they found a weak anti-tumor response after ACT of the CD39 negative fraction (Kortekaas et al., 2020). However, in a recent study, a population of tumor-reactive TILs enriched for a CD39$^-$ CD69$^-$ pool was expanded after adoptive T-cell therapy (Krishna et al., 2020). These results suggest that a pool of tumor antigen-specific T cells are enriched in the CD39 CD69 double negative fraction and can mediate tumor control and T cell persistence.

Recent publications from the group of Dr. Rosenberg have shown that the use of surrogate markers (PD-1) identifies a repertoire of reactive anti-tumor T cells found in the peripheral blood of patients with advanced melanoma (Cafri et al., 2019; Gros et al., 2016, 2019). These findings provide a novel, non-invasive strategy to study and exploit the immune activity of tumor-specific T cells. Potentially providing a robust and scalable strategy to develop highly personalized T-cell products for patients.

6. T-cell differentiation state of TILs

Studies in mouse models have emphasized the importance of the differentiation state of the infused cells (Gattinoni et al., 2011; Restifo et al., 2012). A high positive correlation was found between the persistence of the transferred TILs in the circulation of patients at 1 month and with the induction of partial and complete clinical responses (Rosenberg et al., 2011). These results were confirmed after the characterization of the TCR repertoire of persistent T-cell clones of patients with a clinical benefit after adoptive cell therapy of autologous TILs (Dudley et al., 2002a; Robbins et al., 2004; Zhou et al., 2004, 2005b). These observations lead to further consideration of the mechanisms involved with maintaining T-cell persistence. A limitation for T-cell expansion and functionality is related to the differentiation state of the adoptively transferred T-cell populations. The extent of differentiation is determined by the strength of the TCR signal and the cytokine environment that the T cell encounters during antigen activation. Activated T cells follow a progressive pathway of differentiation from naive T cells into central memory T cell (T_{CM}) and effector memory T cell (T_{EM}) population. Experiments involving T cells expressing transgenic TCRs, showed that a more-differentiated effector T-cell population was less effective for tumor growth control compared to naïve T cells (Gattinoni et al., 2005b). T-cell differentiation is associated with increased production of granzymes and IFNγ, but also with the loss of the ability to produce IL-2, to home to lymph nodes and forced into a proapoptotic senescent state. As the antigen tumor-specific T-cell pool is antigen experienced, they are mainly found in the terminally differentiated T_{EM} repertoire. The in vitro expansion conditions currently used for preparing TILs for ACT—2 weeks or more of culture with high doses of IL-2, anti-CD3 antibody—may cause exaggerated differentiation of isolated TILs. It has been shown that less-differentiated T_{CM} exhibited a potent in vivo recall response when combined with tumor-antigen vaccination and exogenous IL-2, leading to the eradication of large established tumors, compared to T_{EM} (Klebanoff et al., 2005). It has been suggested that cells derived from T_{CM} cells can persist efficiently because of the reacquisition of CD62L. Evidences of limited clinical benefit were observed after ACT of human tumor-reactive CD8 T cell clones that were generated and expanded ex vivo through multiple stimulations (Dudley et al., 2002b). A combined adoptive cell therapy of T_{EM} with T_{CM} mediates a strong reduction of tumor growth in mice compared to single treatments, potentially due to complementary cell killing patterns and local production of IL-2 by T_{EM} (Contreras et al., 2018). These results suggest that T_{CM} and T_{EM} have complementary physiological roles in anti-tumor responses.

More recently, there has been interest in a newly population of CD8 T memory stem cells (T_{SCM}). T_{SCM} cells are identified as the least differentiated and long-lived T-cell memory pool with a strong capacity for self-renewal and a multipotent differentiation into other subsets of memory cells (Gattinoni et al., 2011). T_{SCM} cells resemble naive T cells, as they express the IL-7 receptor α-chain, which can facilitate their survival, as well as high levels of molecules that facilitate their homing to lymph nodes, such as CD62L and CC-chemokine receptor 7 (CCR7). Co-stimulatory receptors such as CD27 and CD28, are highly expressed on T_{SCM} cells, promoting T_{SCM} cells secreting cytokines. Recent studies showed that a stem-like T-cell pool negative for CD39 and CD69 can mediate response of adoptive cell immunotherapy against human cancer (Krishna et al., 2020). The authors suggest that this TIL subset is enriched for tumor-neoantigen T-cell clones capable of limiting tumor growth and generating a pool of effector T cells.

Analysis of TIL clones after ACT, indicates preferential survival of the clonotypes expressing the highest levels of CD28 and CD27, implicating a survival advantage for transferred T cells with an early effector phenotype. The engagement of CD28 with CD80 and/or CD86 on antigen presenting cells amplifies T cell receptor (TCR)-mediated T-cell activation, proliferation, secretion of IL-2, induction of anti-apoptotic molecules and accelerated cell-cycle progression (Acuto and Michel, 2003; Topp et al., 2003). CD27, like CD28, can also augment TCR-induced T-cell proliferation and is required for the generation and maintenance of memory T cells (Arens et al., 2004; Hendriks et al., 2003). One clinical study showed a strong correlation between expression of the phenotypic marker CD27 and clinical response (Powell et al., 2005).

As the memory T-cell pool plays important roles in antitumor ACT-based strategies, several approaches might be developed to promote the phenotypes and functions of memory T cells based on the mechanisms underlying natural memory T-cell formation.

However, IL-2 can also lead to activation-induced cell death (AICD), progressive differentiation (i.e., less "young" TIL) and to the induction of suppressive regulatory T cells However, IL-2 can also lead to activation-induced cell death (AICD), progressive differentiation (i.e., less "young" TIL) and to the induction of suppressive regulatory T cells IL-2 has been the key cytokine applied in the expansion of TIL for ACT. However, the use of IL-2 has also led to collateral expansion of regulatory T cells (Tregs) and progressive T cell differentiation, factors known to limit in vivo persistence and activity of transferred T IL-2 has been the key cytokine applied in the expansion of TIL for ACT. However, the use of IL-2 has also led to collateral expansion of regulatory T cells (Tregs) and progressive T cell differentiation, factors known to limit in vivo persistence and activity of transferred T IL-2 has been the key cytokine used for the in vitro and in vivo expansion of TILs for ACT, due to the signaling through the IL-2Rα promoting T-cell activation, proliferation and survival (Rochman et al., 2009). However, the use of high doses of IL-2 had led to collateral expansion of regulatory T cells and progressive T-cell differentiation toward a short-lived pool (Ahmadzadeh and Rosenberg, 2006; Cesana et al., 2006; Ku et al., 2000). Other cytokines have been proposed for maintaining survival and anti-tumor activities. An in vitro system based on IL-7 and IL-15 that leads to the generation of T_{CM}-like CD8 T cells has been developed with anti-tumor capacity after ACT to tumor-bearing mice (Gattinoni et al., 2005b; Klebanoff et al., 2005; Yuan et al., 2014). IL-21 is a cytokine that shares a common gamma chain with IL-2 and IL-15, and tumor-reactive T cells generated with IL-21 showed a superior anti-tumor effect in vivo compared to T cells grown in other γc cytokines (Hinrichs et al., 2008; Ku et al., 2000; Moroz et al., 2004; Zeng et al., 2005). Recent studies showed that IL-21 also can promote the generation of T_{SCM} cells through the activation of the Janus kinase signal transducer and activator of transcription 3 pathway with superior anti-tumor activity (Alvarez-Fernández et al., 2016).

In summary, ACT of T cells with a naïve, T_{SCM} or T_{CM} profile have shown superior in vivo efficacy in preclinical testing. Thus, methods that selectively expand and maintain a memory-like phenotype T-cell pool are needed. The activation of the Wnt pathway preserves the naive T-cell phenotype, maintains the stemness in memory $CD8^+$ T cells, and arrests effector T-cell differentiation (Muralidharan et al., 2011). It has been shown that reinforcing β-catenin accumulation in cytoplasm and translocation into nucleus with an inhibitor of serine–threonine kinase glycogen synthase kinase-3β (Gsk-3β) or with the Wnt protein family member Wnt3a, promotes the expression of TCF-1 and LEF-1 in the nucleus by binding to β-catenin, keeping T cells in an intermediate state of T_{SCM} (Gattinoni et al., 2009). By blocking T-cell differentiation, the new pool of T cells maintained the self-renewing multipotent functions with proliferative and antitumor capacities.

Clinical methods for rejuvenating the pool of TILs to be transferred will result in longer persistence in the patient and better clinical outcomes. However, the majority of TIL subsets showing tumor reactivity are likely terminally differentiated with a relatively poor proliferative potential. So, strategies aimed at the isolation and expansion of stem-like neoantigen-specific T cells might provide opportunities for future development of more effective T-cell based immunotherapies.

7. Combinatorial therapies

The possibility to combine immune checkpoint inhibitors (ICIs) with ACT is an appealing therapeutic approach that is currently being explored. The activity of TILs can be hindered by antigen-presenting cells or tumor cells through immune checkpoints such as the PD-1/PD-L1 axis or the CTLA-4/CD80/CD86 axis. Manipulating co-inhibitory pathways may improve the anti-tumor functions of the TIL infused product.

Several studies have shown overexpression of PD-1 and CLTA-4 on in vitro expanded TILs, indicating that the use of PD-1 or CTLA-4 inhibitors could be a relevant combination strategy to prevent immune-mediated inactivation in vivo (Bjoern et al., 2017; Donia et al., 2017; Forget et al., 2018; Mullinax et al., 2018; van den Berg et al., 2020; Zippel et al., 2019). Prospective analyses of clinical trials based on ICIs treatment have shown that a subset of T cells is reinvigorated in responder patients (Friese et al., 2020; Kurtulus et al., 2019; Miller et al., 2019; Sade-Feldman et al., 2018; Siddiqui et al., 2019). The pool of $TCF1^+PD-1^+$ cells were detected among tumor-reactive CD8 T cells in the blood of melanoma patients and among TILs resected from primary melanomas. Same results were found in preclinical mice models (Siddiqui et al., 2019) and chronic viral infections(Im et al., 2016; Utzschneider et al., 2016).

Several ongoing clinical trials are studying the safety, side effects, and benefits of combining the administration of tumor-infiltrating lymphocytes (TILs) with several ICIs (Table 1). One study found that combined TILs and anti-PD1 therapy significantly increased ORR, PFS and OS of osteosarcoma patients that exhibited disease progression after first-line therapy relative to anti-PD1 therapy alone (Wang et al., 2020). In another study, the combination of adoptive TILs and anti-PD1 inhibitor in patients with chemotherapy-resistant cervical cancer resulted in a clinical benefit

(Yin et al., 2020). A TIL-based ACT with ICIs was evaluated in a clinical phase I/II trial (NCT03296137) (Kverneland et al., 2021). Ipilimumab was administered prior to tumor resection and nivolumab was administered in combination with the TIL infusion product. Objective response, safety and feasibility were comparable to other trials of ACT on metastatic melanoma with the addition of expected ICIs toxicity.

Although checkpoint blockade has improved survival in a subset of patients, the actual clinical challenge is to find alternative therapies which can improve the survival of patients who are checkpoint resistant or refractory (Hodi et al., 2018). There is no clear standard of care therapy for patients with checkpoint inhibitor refractory disease. Thus, adoptive cell transfer of TILs has been evaluated for this purpose. In a recent study, the infusion of the TIL-based ACT product of Lifileucel produced by Iovance Biotherapeutics, was shown to produce a durable response against metastatic melanoma in large cohorts of anti-PD-1/PD-L1 refractory patients. However, in another study, they found lower objective response rates and overall survival in patients whose tumors were refractory to anti–PD-1 or BRAF/MEK inhibitors (Seitter et al., 2021). Similar results were obtained for CLTA-4 ICIs, in which patients that were exposed to an anti-CTLA-4 inhibitor prior to TILs therapy experienced a clinical response of shorter duration (Forget et al., 2018).

Larger trials are required to further define the optimal biomarkers of response and the efficacy of TILs from patients who have progressed on previous ICIs-based treatments.

8. Synthetic TILs in cancer therapy

In spite of the tremendous progress achieved in TIL-based therapy, a high proportion of melanoma patients treated with ACT-TIL strategy are non-responders. One of the many hurdles facing the use of TILs is the limited anti-tumor capacity of TIL-based immunotherapy. So, in order to improve cytotoxicity of TILs, a sustained expression of adjuvant cytokines may enhance T-cell reactivity toward cancer cells.

In a first approach, a pool of TILs was genetically engineered to secrete IL-2 (Heemskerk et al., 2008). The in vitro results were promising, as six out of the eight transduced patient samples produced IL-2 upon autologous tumor stimulation and persisted longer than non-transduced TILs. Unfortunately, the clinical trial showed a reduced clinical benefit compared to previous trials with non-genetically modified TILs. In another study,

TILs were genetically manipulated to express an inducible single-chain IL-12 gene placed under the control of a nuclear factor of activated T cells-(NFAT)-responsive promoter (Zhang et al., 2015). Results from the clinical trial showed that 63% of patients exhibited an objective clinical response. Nevertheless, the clinical benefit was short and IL-12-producing cells rarely persist at one-month post-infusion. Moreover, high toxicities were found in patients receiving increasing doses of the TIL product. Even though the production of IL-12 was regulated by the NFAT inducible promoter, high serum IL-12 and IFNγ levels indicated that TILs were being activated outside of the tumor microenvironment, which was likely a contributing factor to the toxicities. More recently, a transient expression of IL-12 by mRNA electroporation of mouse TILs showed that intratumoral adoptive transfer of these cells, alone or in combination with an CD137 agonist agent, may be a highly effective strategy for clearing tumors (Etxeberria et al., 2019).

Another limiting factor in ACT-TIL based strategies is the insufficient homing of T cells to the tumor niche. Several research groups have evaluated the improvement of tumor trafficking through the overexpression of chemokines receptors (Kershaw et al., 2002). In several cancer models, CXCL1 and CXCL8 have been shown to be produced by cells within the tumor microenvironment (TME), so gene transfer of CXCR2 DNA into TILs, may improve adoptive cell therapies and limit off-target toxicities. Transgenic TILs carrying a MAGE-A3 high affinity TCR with CXCR2 increased tumor infiltration and limited tumor growth in a melanoma mice model (Idorn et al., 2018; Kershaw et al., 2002). The positive results obtained in preclinical models allowed the design a clinical trial in humans. A phase I/II study (NCT01740557) aimed to evaluate feasibility and safety of CXCR2 and NGFR transduced tumor-infiltrating lymphocytes for treating metastatic malignant melanoma. Another objective of this clinical trial was to determine whether CXCR2 transduction enhances the ability of TILs to migrate to melanoma tumors, and assess its association with the clinical response. However, results from this pilot clinical trial have not been reported yet.

After chronic antigen exposure, T cells express coinhibitory receptors, named as markers of T-cell exhaustion, to reduce the antitumor activities of TILs. Some groups have attempted to genetically eliminate the immunosuppressive and exhaustive inhibitory receptors by genetic modification of TILs. Gene-editing strategies for PD-1 disruption have been studied, such as transcription activator-like effector nucleases (TALENs) (Menger et al., 2016), zinc finger nucleases (ZFNs) (Beane et al., 2015) and CRISPR/Cas9 system (Su et al., 2016). In vitro experiments prove that a permanent disruption of the PD-1 gene locus exhibited a polyfunctional cytokine secretion profile with an up-regulation of IFNγ production and enhanced cytotoxicity when cocultured with a tumor cell line (Beane et al., 2015; Menger et al., 2016; Su et al., 2016). Furthermore, the adoptive cell transfer of tumor-reactive CD8 T cells modified to downregulate the expression of PD-1 enhanced the persistence at the tumor site and increased tumor control (Menger et al., 2016). Even though, some previous works showed that the lack of PD-1/PD-L1 signals during some primary infections resulted in more robust effector T-cell responses, a permanent absence of PD-1 promotes the accumulation of terminally differentiated exhausted or senescent CD8 T cells (Odorizzi et al., 2015; Otano et al., 2018).

A recent clinical trial was initiated following a ground-breaking research in which CRISPR-Cas9-mediated the inactivation of the cytokine-inducible SH2 (CISH) gene in TILs, increased their capability to control solid tumors (NCT04426669). CISH is a novel intra-cellular immune checkpoint that negatively regulates TCR signaling (Palmer et al., 2015). The adoptive transfer of genetically modified T cells for CISH show a durable tumor regression and survival in a preclinical animal model. The observed clinical benefit is intensified when combined with an anti-PD1 antibody (Palmer et al., 2020).

9. Perspectives and conclusions

ACT-TIL is a promising emerging immunotherapy for metastatic melanoma patients that is expanding to other epithelial malignancies. The innovative application of the advances and discoveries in other fields continues to shape and improve this evolving therapy However, there are a number of challenges associated with the production and delivery of these therapies. TIL-ACT is an ultimate personalized treatment since a specific infusion product has to be manufactured for each individual patient. This requires a selection of eligible patients for TIL treatment and a highly specialized good manufacturing practice (GMP).

Further pre-clinical and clinical studies are needed for future modifications to the therapy, the forthcoming future of TIL-based ACT will likely be highly personalized, interdisciplinary, and an effective treatment option in a wide variety of cancers. Moreover, the future of ACT will rely on optimal combinatorial regimens to augment clinical efficacy and overcome resistance mechanisms.

Acknowledgments

I.O. is supported by the AECC Investigator 2020 and I.D.C. is supported by the AECC PhD fellowship 2022.

Author contributions

Conception and design, I.O., L.P.A.; manuscript writing, I.O., L.P.A., I.D.C.; and writing review and editing, I.O., L.P.A., I.D.C. All authors performed a critical revision of the manuscript for important intellectual content and final approval of the manuscript.

Competing interests

L.P.A. reports grants or contracts from Merck Sharp & Dohme, AstraZeneca, Pfizer, Bristol Myers Squibb; consulting fees from Lilly, Merck Sharp & Dohme, Roche, Pharmamar, Merck, AstraZeneca, Novartis, Servier, Amgen, Pfizer, Ipsen, Sanofi, Bayer, Blueprint, Bristol Myers Squibb, Mirati; payment or honoraria for lectures, presentations, speakers bureaus, manuscript writing or educational events from AstraZeneca, Janssen, Merck, Mirati, Sanofi; and leadership or fiduciary role in other board, society, committee or advocacy group, paid or unpaid, for Genomica and Altum sequency.

References

Acuto, O., Michel, F., 2003. CD28-mediated co-stimulation: a quantitative support for TCR signalling. Nat. Rev. Immunol. 3.

Ahmadzadeh, M., Rosenberg, S.A., 2006. IL-2 administration increases CD4+CD25hi Foxp3 + regulatory T cells in cancer patients. Blood 107.

Ahrends, T., et al., 2017. CD4+ T cell help confers a cytotoxic T cell effector program including coinhibitory receptor downregulation and increased tissue invasiveness. Immunity 47.

Ahrends, T., et al., 2019. CD4+ T cell help creates memory CD8+ T cells with innate and help-independent recall capacities. Nat. Commun. 10.

Alvarez-Fernández, C., Escribà-Garcia, L., Vidal, S., Sierra, J., Briones, J., 2016. A short CD3/CD28 costimulation combined with IL-21 enhance the generation of human memory stem T cells for adoptive immunotherapy. J. Transl. Med. 14.

Anders, K., et al., 2011. Oncogene-targeting T cells reject large tumors while oncogene inactivation selects escape variants in mouse models of cancer. Cancer Cell 20.

Andersen, R.S., et al., 2012. Dissection of T-cell antigen specificity in human melanoma. Cancer Res. 72.

Andersen, R., et al., 2016. Long-lasting complete responses in patients with metastatic melanoma after adoptive cell therapy with tumor-infiltrating lymphocytes and an attenuated il2 regimen. Clin. Cancer Res. 22.

Andersen, R., et al., 2018. T-cell responses in the microenvironment of primary renal cell carcinoma-implications for adoptive cell therapy. Cancer Immunol. Res. 6.

Arens, R., et al., 2004. Tumor rejection induced by CD70-mediated quantitative and qualitative effects on effector CD8+ T cell formation. J. Exp. Med. 199.

Beane, J.D., et al., 2015. Clinical scale zinc finger nuclease-mediated gene editing of PD-1 in tumor infiltrating lymphocytes for the treatment of metastatic melanoma. Mol. Ther. 23.

Besser, M.J., et al., 2013. Adoptive transfer of tumor-infiltrating lymphocytes in patients with metastatic melanoma: intent-to-treat analysis and efficacy after failure to prior immunotherapies. Clin. Cancer Res. 19.

Bjoern, J., et al., 2017. Influence of ipilimumab on expanded tumour derived T cells from patients with metastatic melanoma. Oncotarget 8.

Borst, J., Ahrends, T., Bąbała, N., Melief, C.J.M., Kastenmüller, W., 2018. CD4+ T cell help in cancer immunology and immunotherapy. Nat. Rev. Immunol. 18.

Buonaguro, L., Petrizzo, A., Tornesello, M.L., Buonaguro, F.M., 2011. Translating tumor antigens into cancer vaccines. Clin. Vaccine Immunol. 18.

Cafri, G., et al., 2019. Memory T cells targeting oncogenic mutations detected in peripheral blood of epithelial cancer patients. Nat. Commun. 10.

Castle, J.C., et al., 2012. Exploiting the mutanome for tumor vaccination. Cancer Res. 72.

Cesana, G.C., et al., 2006. Characterization of CD4+CD25+ regulatory T cells in patients treated with high-dose interleukin-2 for metastatic melanoma or renal cell carcinoma. J. Clin. Oncol. 24.

Chacon, J.A., Sarnaik, A.A., Pilon-Thomas, S., Radvanyi, L., 2015a. Triggering co-stimulation directly in melanoma tumor fragments drives CD8+ tumor-infiltrating lymphocyte expansion with improved effector-memory properties. OncoImmunology 4.

Chacon, J.A., et al., 2015b. Manipulating the tumor microenvironment ex vivo for enhanced expansion of tumor-infiltrating lymphocytes for adoptive cell therapy. Clin. Cancer Res. 21.

Chapuis, A.G., et al., 2012. Transferred melanoma-specific CD8+ T cells persist, mediate tumor regression, and acquire central memory phenotype. Proc. Natl. Acad. Sci. U. S. A. 109.

Chapuis, A.G., et al., 2013. Transferred WT1-reactive CD8+ T cells can mediate antileukemic activity and persist in post-transplant patients. Sci. Transl. Med. 5.

Chapuis, A.G., et al., 2014. Regression of metastatic Merkel cell carcinoma following transfer of polyomavirus-specific T cells and therapies capable of re-inducing HLA class-I. Cancer Immunol. Res. 2.

Chen, Y.T., et al., 1997. A testicular antigen aberrantly expressed in human cancers detected by autologous antibody screening. Proc. Natl. Acad. Sci. U. S. A. 94.

Ciriello, G., et al., 2013. Emerging landscape of oncogenic signatures across human cancers. Nat. Genet. 45.

Contreras, A., et al., 2018. Co-transfer of tumor-specific effector and memory CD8+ T cells enhances the efficacy of adoptive melanoma immunotherapy in a mouse model. J. Immunother. Cancer 6.

Coulie, P.G., van den Eynde, B.J., van der Bruggen, P., Boon, T., 2014. Tumour antigens recognized by T lymphocytes: at the core of cancer immunotherapy. Nat. Rev. Cancer 14.

Danahy, D.B., Berton, R.R., Badovinac, V.P., 2020. Cutting edge: antitumor immunity by pathogen-specific CD8 T cells in the absence of cognate antigen recognition. J. Immunol. 204.

Delorme, E.J., Alexander, P., 1964. Treatment of primary fibrosarcoma in the rat with immune lymphocytes. The Lancet 284.

Donia, M., et al., 2012. Characterization and comparison of "standard" and "young" tumour-infiltrating lymphocytes for adoptive cell therapy at a Danish translational research institution. Scand. J. Immunol. 75.

Donia, M., et al., 2017. PD-1þ polyfunctional T cells dominate the periphery after tumor-infiltrating lymphocyte therapy for cancer. Clin. Cancer Res. 23.

Dudley, M.E., et al., 2002a. Cancer regression and autoimmunity in patients after clonal repopulation with antitumor lymphocytes. Science 298.

Dudley, M.E., et al., 2002b. A phase I study of nonmyeloablative chemotherapy and adoptive transfer of autologous tumor antigen-specific T lymphocytes in patients with metastatic melanoma. J. Immunother. 25.

Dudley, M.E., et al., 2008. Adoptive cell therapy for patients with metastatic melanoma: evaluation of intensive myeloablative chemoradiation preparative regimens. J. Clin. Oncol. 26.

Dudley, M.E., et al., 2010. CD8+ enriched "Young" tumor infiltrating lymphocytes can mediate regression of metastatic melanoma. Clin. Cancer Res. 16.

Duhen, T., et al., 2018. Co-expression of CD39 and CD103 identifies tumor-reactive CD8 T cells in human solid tumors. Nat. Commun. 9.

Etxeberria, I., et al., 2019. Intratumor adoptive transfer of IL-12 mRNA transiently engineered antitumor CD8$^+$ T cells. Cancer Cell 36.

Fefer, A., 1969. Immunotherapy and chemotherapy of moloney sarcoma virus-induced tumors in mice. Cancer Res. 29.

Feng, H., Shuda, M., Chang, Y., Moore, P.S., 2008. Clonal integration of a polyomavirus in human Merkel cell carcinoma. Science 319.

Fernandez-Poma, S.M., et al., 2017. Expansion of tumor-infiltrating CD8+ T cells expressing PD-1 improves the efficacy of adoptive T-cell therapy. Cancer Res. 77, 3672–3684.

Filatenkov, A., et al., 2015. Ablative tumor radiation can change the tumor immune cell microenvironment to induce durable complete remissions. Clin. Cancer Res. 21.

Forget, M.A., et al., 2018. Prospective analysis of adoptive TIL therapy in patients with metastatic melanoma: response, impact of anti-CTLA4, and biomarkers to predict clinical outcome. Clin. Cancer Res. 24.

Freedman, R.S., et al., 1994. Intraperitoneal adoptive immunotherapy of ovarian carcinoma with tumor-infiltrating lymphocytes and low-dose recombinant interleukin-2: a pilot trial. J. Immunother. 16.

Friese, C., et al., 2020. CTLA-4 blockade boosts the expansion of tumor-reactive CD8+ tumor-infiltrating lymphocytes in ovarian cancer. Sci. Rep. 10.

Fujita, K., et al., 1995. Prolonged disease-free period in patients with advanced epithelial ovarian cancer after adoptive transfer of tumor-infiltrating lymphocytes. Clin. Cancer Res. 1.

Gartner, J.J., et al., 2021. A machine learning model for ranking candidate HLA class I neoantigens based on known neoepitopes from multiple human tumor types. Nat. Cancer 2.

Gattinoni, L., et al., 2005a. Removal of homeostatic cytokine sinks by lymphodepletion enhances the efficacy of adoptively transferred tumor-specific CD8+ T cells. J. Exp. Med. 202.

Gattinoni, L., et al., 2005b. Acquisition of full effector function in vitro paradoxically impairs the in vivo antitumor efficacy of adoptively transferred CD8+ T cells. J. Clin. Investig. 115.

Gattinoni, L., et al., 2009. Wnt signaling arrests effector T cell differentiation and generates CD8+ memory stem cells. Nat. Med. 15.

Gattinoni, L., et al., 2011. A human memory T cell subset with stem cell-like properties. Nat. Med. 17.

Goff, S.L., et al., 2016. Randomized, prospective evaluation comparing intensity of lymphodepletion before adoptive transfer of tumor-infiltrating lymphocytes for patients with metastatic melanoma. J. Clin. Oncol. 34.

Gros, A., et al., 2014. PD-1 identifies the patient-specific CD8+ tumor-reactive repertoire infiltrating human tumors. J. Clin. Investig. 124.

Gros, A., et al., 2016. Prospective identification of neoantigen-specific lymphocytes in the peripheral blood of melanoma patients. Nat. Med. 22.

Gros, A., et al., 2019. Recognition of human gastrointestinal cancer neoantigens by circulating PD-1+ lymphocytes. J. Clin. Investig. 129.

Gross, L., 1943. Intradermal immunization of C3H mice against a sarcoma that originated in an animal of the same line. Cancer Res. 3.

Gubin, M.M., et al., 2014. Checkpoint blockade cancer immunotherapy targets tumour-specific mutant antigens. Nature 515.

Gubin, M.M., Artyomov, M.N., Mardis, E.R., Schreiber, R.D., 2015. Tumor neoantigens: building a framework for personalized cancer immunotherapy. J. Clin. Investig. 125.

Gupta, P.K., et al., 2015. CD39 expression identifies terminally exhausted CD8+ T cells. PLoS Pathog. 11.

Heemskerk, B., et al., 2008. Adoptive cell therapy for patients with melanoma, using tumor-infiltrating lymphocytes genetically engineered to secrete interleukin-2. Hum. Gene Ther. 19.

Hendriks, J., Xiao, Y., Borst, J., 2003. CD27 promotes survival of activated T cells and complements CD28 in generation and establishment of the effector T cell pool. J. Exp. Med. 198.

Heslop, H.E., et al., 2010. Long-term outcome of EBV-specific T-cell infusions to prevent or treat EBV-related lymphoproliferative disease in transplant recipients. Blood 115.

Hinrichs, C.S., et al., 2008. IL-2 and IL-21 confer opposing differentiation programs to CD8 + T cells for adoptive immunotherapy. Blood 111.

Hodi, F.S., et al., 2018. Nivolumab plus ipilimumab or nivolumab alone versus ipilimumab alone in advanced melanoma (CheckMate 067): 4-year outcomes of a multicentre, randomised, phase 3 trial. Lancet Oncol. 19.

Huang, J., et al., 2004. T cells associated with tumor regression recognize frameshifted products of the CDKN2A tumor suppressor gene locus and a mutated HLA class I gene product. J. Immunol. 172.

Hugo, W., et al., 2016. Genomic and transcriptomic features of response to anti-PD-1 therapy in metastatic melanoma. Cell 165.

Hunder, N.N., et al., 2008. Treatment of metastatic melanoma with autologous CD4+ T cells against NY-ESO-1. N. Engl. J. Med. 358.

Idorn, M., et al., 2018. Chemokine receptor engineering of T cells with CXCR2 improves homing towards subcutaneous human melanomas in xenograft mouse model. OncoImmunology 7.

Im, S.J., et al., 2016. Defining CD8+ T cells that provide the proliferative burst after PD-1 therapy. Nature 537.

Itzhaki, O., et al., 2011. Establishment and large-scale expansion of minimally cultured young tumor infiltrating lymphocytes for adoptive transfer therapy. J. Immunother. 34.

Jazaeri, A.A., et al., 2019. Safety and efficacy of adoptive cell transfer using autologous tumor infiltrating lymphocytes (LN-145) for treatment of recurrent, metastatic, or persistent cervical carcinoma. J. Clin. Oncol. 37.

Jiang, S.S., et al., 2015. A phase I clinical trial utilizing autologous tumor-infiltrating lymphocytes in patients with primary hepatocellular carcinoma. Oncotarget 6.

Jing, W., et al., 2017. Adoptive cell therapy using PD-1+ myeloma-reactive T cells eliminates established myeloma in mice. J. Immunother. Cancer 5.

Kansas, G.S., Wood, G.S., Tedder, T.F., 1991. Expression, distribution, and biochemistry of human CD39. Role in activation-associated homotypic adhesion of lymphocytes. J. Immunol. (Balt. Md. 1950) 146.

Kenter, G.G., et al., 2009. Vaccination against HPV-16 oncoproteins for vulvar intraepithelial neoplasia. N. Engl. J. Med. 361.

Kershaw, M.H., et al., 2002. Redirecting migration of T cells to chemokine secreted from tumors by genetic modification with CXCR2. Hum. Gene Ther. 13.

Klebanoff, C.A., et al., 2005. Central memory self/tumor-reactive CD8+ T cells confer superior antitumor immunity compared with effector memory T cells. Proc. Natl. Acad. Sci. U. S. A. 102.

Kortekaas, K.E., et al., 2020. CD39 identifies the CD4þ tumor-specific T-cell population in human cancer. Cancer Immunol. Res. 8.

Kozin, S.V., et al., 2010. Recruitment of myeloid but not endothelial precursor cells facilitates tumor regrowth after local irradiation. Cancer Res. 70.

Kradin, R.L., et al., 1987. Tumor-derived interleukin-2-dependent lymphocytes in adoptive immunotherapy of lung cancer. Cancer Immunol. Immunother. 24.

Kradin, R.L., et al., 1989. Tumour-infiltrating lymphocytes and INTERLEUKIN-2 in treatment of advanced cancer. The Lancet 333.

Kreiter, S., et al., 2015. Mutant MHC class II epitopes drive therapeutic immune responses to cancer. Nature 520.

Krishna, S., et al., 2020. Stem-like CD8 T cells mediate response of adoptive cell immunotherapy against human cancer. Science 370.

Kristensen, N.P., et al., 2022. Neoantigen-reactive CD8+ T cells affect clinical outcome of adoptive cell therapy with tumor-infiltrating lymphocytes in melanoma. J. Clin. Investig. 132.

Ku, C.C., Murakami, M., Sakamoto, A., Kappler, J., Marrack, P., 2000. Control of homeostasis of CD8+ memory T cells by opposing cytokines. Science 288.

Kumar, A., Watkins, R., Vilgelm, A.E., 2021. Cell therapy with TILs: training and taming T cells to fight cancer. Front. Immunol. 12.

Kurtulus, S., et al., 2019. Checkpoint blockade immunotherapy induces dynamic changes in PD-1–CD8+ tumor-infiltrating T cells. Immunity 50.

Kverneland, A.H., et al., 2020. Adoptive cell therapy in combination with checkpoint inhibitors in ovarian cancer. Oncotarget 11.

Kverneland, A.H., et al., 2021. Adoptive cell therapy with tumor-infiltrating lymphocytes supported by checkpoint inhibition across multiple solid cancer types. J. Immunother. Cancer 9.

Lawrence, M.S., et al., 2013. Mutational heterogeneity in cancer and the search for new cancer-associated genes. Nature 499.

Le, D.T., et al., 2017. Mismatch repair deficiency predicts response of solid tumors to PD-1 blockade. Science 357.

Lee, H.J., et al., 2017. Expansion of tumor-infiltrating lymphocytes and their potential for application as adoptive cell transfer therapy in human breast cancer. Oncotarget 8.

Leem, G., et al., 2020. 4-1BB co-stimulation further enhances anti-PD-1-mediated reinvigoration of exhausted CD39+CD8 T cells from primary and metastatic sites of epithelial ovarian cancers. J. Immunother. Cancer 8.

Linette, G.P., Carreno, B.M., 2019. Tumor-infiltrating lymphocytes in the checkpoint inhibitor era. Curr. Hematol. Malig. Rep. 14.

Liu, Z., et al., 2020. Detecting tumor antigen-specific T cells via interaction-dependent fucosyl-biotinylation. Cell 183.

Lu, Y.-C., et al., 2013. Mutated PPP1R3B is recognized by T cells used to treat a melanoma patient who experienced a durable complete tumor regression. J. Immunol. 190.

Lu, Y.C., et al., 2014. Efficient identification of mutated cancer antigens recognized by T cells associated with durable tumor regressions. Clin. Cancer Res. 20.

Matsushita, H., et al., 2012. Cancer exome analysis reveals a T-cell-dependent mechanism of cancer immunoediting. Nature 482.

Meier, S.L., Satpathy, A.T., Wells, D.K., 2022. Bystander T cells in cancer immunology and therapy. Nat. Cancer 3, 143–155.

Menger, L., et al., 2016. TALEN-mediated inactivation of PD-1 in tumor-reactive lymphocytes promotes intratumoral T-cell persistence and rejection of established tumors. Cancer Res. 76.

Miller, B.C., et al., 2019. Subsets of exhausted CD8+ T cells differentially mediate tumor control and respond to checkpoint blockade. Nat. Immunol. 20.

Moesta, A.K., Li, X.Y., Smyth, M.J., 2020. Targeting CD39 in cancer. Nat. Rev. Immunol. 20.

Moroz, A., et al., 2004. IL-21 enhances and sustains CD8 + T cell responses to achieve durable tumor immunity: comparative evaluation of IL-2, IL-15, and IL-21. J. Immunol. 173.

Mullinax, J.E., et al., 2018. Combination of ipilimumab and adoptive cell therapy with tumor-infiltrating lymphocytes for patients with metastatic melanoma. Front. Oncol. 8.

Muralidharan, S., et al., 2011. Activation of Wnt signaling arrests effector differentiation in human peripheral and cord blood-derived T lymphocytes. J. Immunol. 187.

Muul, L.M., Spiess, P.J., Director, E.P., Rosenberg, S.A., 1987. Identification of specific cytolytic immune responses against autologous tumor in humans bearing malignant melanoma. J. Immunol. (Balt. Md. 1950) 138.

Nelson, M.H., et al., 2016. Toll-like receptor agonist therapy can profoundly augment the antitumor activity of adoptively transferred CD8 + T cells without host preconditioning. J. Immunother. Cancer 4.

Odorizzi, P.M., Pauken, K.E., Paley, M.A., Sharpe, A., John Wherry, E., 2015. Genetic absence of PD-1 promotes accumulation of terminally differentiated exhausted CD8 + T cells. J. Exp. Med. 212.

Otano, I., et al., 2018. Molecular recalibration of PD-1 + antigen-specific T cells from blood and liver. Mol. Ther. 26, 2553–2566.

Palmer, D.C., et al., 2015. Cish actively silences TCR signaling in CD8 + T cells to maintain tumor tolerance. J. Exp. Med. 212.

Palmer, D., et al., 2020. 333 Targeting the apical intracellular checkpoint CISH unleashes T cell neoantigen reactivity and effector program. J. Immunother. Cancer 8.

Paria, B.C., et al., 2021. Rapid identification and evaluation of neoantigen-reactive T-cell receptors from single cells. J. Immunother. 44.

Parkhurst, M., et al., 2017. Isolation of T-cell receptors specifically reactive with mutated tumor-associated antigens from tumor-infiltrating lymphocytes based on CD137 expression. Clin. Cancer Res. 23.

Pilon-Thomas, S., et al., 2012. Efficacy of adoptive cell transfer of tumor-infiltrating lymphocytes after lymphopenia induction for metastatic melanoma. J. Immunother. 35.

Powell, D.J., Dudley, M.E., Robbins, P.F., Rosenberg, S.A., 2005. Transition of late-stage effector T cells to CD27 + CD28 + tumor-reactive effector memory T cells in humans after adoptive cell transfer therapy. Blood 105.

Prickett, T.D., et al., 2016. Durable complete response from metastatic melanoma after transfer of autologous T cells recognizing 10 mutated tumor antigens. Cancer Immunol. Res. 4.

Radvanyi, L.G., et al., 2012. Specific lymphocyte subsets predict response to adoptive cell therapy using expanded autologous tumor-infiltrating lymphocytes in metastatic melanoma patients. Clin. Cancer Res. 18.

Restifo, N.P., Dudley, M.E., Rosenberg, S.A., 2012. Adoptive immunotherapy for cancer: harnessing the T cell response. Nat. Rev. Immunol. 12.

Robbins, P.F., et al., 2004. Cutting edge: persistence of transferred lymphocyte clonotypes correlates with cancer regression in patients receiving cell transfer therapy. J. Immunol. 173.

Robbins, P.F., et al., 2013. Mining exomic sequencing data to identify mutated antigens recognized by adoptively transferred tumor-reactive T cells. Nat. Med. 19.

Rochman, Y., Spolski, R., Leonard, W.J., 2009. New insights into the regulation of T cells by γc family cytokines. Nat. Rev. Immunol. 9.

Rosenberg, S.A., Restifo, N.P., 2015. Adoptive cell transfer as personalized immunotherapy for human cancer. Science 348.

Rosenberg, S.A., Spiess, P., Lafreniere, R., 1986. A new approach to the adoptive immunotherapy of cancer with tumor-infiltrating lymphocytes. Science 233.

Rosenberg, S.A., et al., 1988. Use of tumor-infiltrating lymphocytes and Interleukin-2 in the immunotherapy of patients with metastatic melanoma. N. Engl. J. Med. 319.

Rosenberg, S.A., et al., 1994. Treatment of patients with metastatic melanoma with autologous tumor-infiltrating lymphocytes and interleukin 2. J. Natl. Cancer Inst. 86.

Rosenberg, S.A., et al., 2011. Durable complete responses in heavily pretreated patients with metastatic melanoma using T-cell transfer immunotherapy. Clin. Cancer Res. 17.

Roudko, V., et al., 2020. Shared immunogenic poly-epitope frameshift mutations in microsatellite unstable tumors. Cell 183.

Sade-Feldman, M., et al., 2018. Defining T cell states associated with response to checkpoint immunotherapy in melanoma. Cell 175.

Salas-Benito, D., et al., 2021. The mutational load and a T-cell inflamed tumour phenotype identify ovarian cancer patients rendering tumour-reactive T cells from PD-1+ tumour-infiltrating lymphocytes. Br. J. Cancer 124.

Sarnaik, A., et al., 2020. Long-term follow up of lifileucel (LN-144) cryopreserved autologous tumor infiltrating lymphocyte therapy in patients with advanced melanoma progressed on multiple prior therapies. J. Clin. Oncol. 38.

Saxena, M., van der Burg, S.H., Melief, C.J.M., Bhardwaj, N., 2021. Therapeutic cancer vaccines. Nat. Rev. Cancer 21.

Scheper, W., et al., 2019. Low and variable tumor reactivity of the intratumoral TCR repertoire in human cancers. Nat. Med. 25.

Schluns, K.S., Kieper, W.C., Jameson, S.C., Lefrançois, L., 2000. Interleukin-7 mediates the homeostasis of naïve and memory CD8 T cells in vivo. Nat. Immunol. 1.

Schluns, K.S., Williams, K., Ma, A., Zheng, X.X., Lefrançois, L., 2002. Cutting edge: requirement for IL-15 in the generation of primary and memory antigen-specific CD8 T cells. J. Immunol. 168.

Seitter, S.J., et al., 2021. Impact of prior treatment on the efficacy of adoptive transfer of tumor-infiltrating lymphocytes in patients with metastatic melanoma. Clin. Cancer Res. 27.

Shuford, W.W., et al., 1997. 4-1BB costimulatory signals preferentially induce CD8+ T cell proliferation and lead to the amplification in vivo of cytotoxic T cell responses. J. Exp. Med. 186.

Siddiqui, I., et al., 2019. Intratumoral Tcf1+ PD-1+ CD8+ T cells with stem-like properties promote tumor control in response to vaccination and checkpoint blockade immunotherapy. Immunity 50.

Simoni, Y., et al., 2018. Bystander CD8+ T cells are abundant and phenotypically distinct in human tumour infiltrates. Nature 557.

Snyder, A., et al., 2014. Genetic basis for clinical response to CTLA-4 blockade in melanoma. N. Engl. J. Med. 371.

Stern, P.L., Harrop, R., 2017. 5T4 oncofoetal antigen: an attractive target for immune intervention in cancer. Cancer Immunol. Immunother. 66, 415–426.

Stevanović, S., et al., 2015. Complete regression of metastatic cervical cancer after treatment with human papillomavirus-targeted tumor-infiltrating T cells. J. Clin. Oncol. 33.

Stevanovic, S., et al., 2019. A phase II study of tumor-infiltrating lymphocyte therapy for human papillomavirus–associated epithelial cancers. Clin. Cancer Res. 25.

Su, S., et al., 2016. CRISPR-Cas9 mediated efficient PD-1 disruption on human primary T cells from cancer patients. Sci. Rep. 6.

Topalian, S.L., et al., 1988. Immunotherapy of patients with advanced cancer using tumor-infiltrating lymphocytes and recombinant interleukin-2: a pilot study. J. Clin. Oncol. 6.

Topp, M.S., et al., 2003. Restoration of CD28 expression in CD28- CD8+ memory effector T cells reconstitutes antigen-induced IL-2 production. J. Exp. Med. 198.

Tran, E., et al., 2016. T-cell transfer therapy targeting mutant KRAS in cancer. N. Engl. J. Med. 375.

Tran, E., et al., 2014. Cancer immunotherapy based on mutation-specific CD4+ T cells in a patient with epithelial cancer. Science 344.

Utzschneider, D.T., et al., 2016. T cell factor 1-expressing memory-like CD8+ T cells sustain the immune response to chronic viral infections. Immunity 45.

van Allen, E.M., et al., 2015. Genomic correlates of response to CTLA-4 blockade in metastatic melanoma. Science 350.

van den Bulk, J., et al., 2019. Neoantigen-specific immunity in low mutation burden colorectal cancers of the consensus molecular subtype 4. Genome Med. 11.

van den Berg, J.H., et al., 2020. Tumor infiltrating lymphocytes (TIL) therapy in metastatic melanoma: boosting of neoantigen-specific T cell reactivity and long-term follow-up. J. Immunother. Cancer 8.

Vincent, J., et al., 2010. 5-Fluorouracil selectively kills tumor-associated myeloid-derived suppressor cells resulting in enhanced T cell-dependent antitumor immunity. Cancer Res. 70.

Watts, T.H., 2005. TNF/TNFR family members in costimulation of T cell responses. Annu. Rev. Immunol. 23.

Wang, X., Rivière, I., 2015. Manufacture of tumor- and virus-specific T lymphocytes for adoptive cell therapies. Cancer Gene Ther. 22.

Wang, C., Li, M., Wei, R., Wu, J., 2020. Adoptive transfer of TILs plus anti-PD1 therapy: an alternative combination therapy for treating metastatic osteosarcoma. J. Bone Oncol. 25.

Weiden, P.L., et al., 1979. Antileukemic effect of graft-versus-host disease in human recipients of allogeneic-marrow grafts. N. Engl. J. Med. 300.

Wells, D.K., et al., 2020. Key parameters of tumor epitope immunogenicity revealed through a consortium approach improve neoantigen prediction. Cell 183.

Wrzesinski, C., et al., 2010. Increased intensity lymphodepletion enhances tumor treatment efficacy of adoptively transferred tumor-specific T cells. J. Immunother. 33.

Yadav, M., et al., 2014. Predicting immunogenic tumour mutations by combining mass spectrometry and exome sequencing. Nature 515.

Ye, Q., et al., 2014. CD137 accurately identifies and enriches for naturally occurring tumor-reactive T cells in tumor. Clin. Cancer Res. 20.

Yin, H., et al., 2020. TILs and anti-PD1 therapy: an alternative combination therapy for PDL1 negative metastatic cervical cancer. J. Immunol. Res. 2020.

Yuan, C.H., et al., 2014. Interleukin-7 enhances the in vivo anti-tumor activity of tumor-reactive CD8+ T cells with induction of IFN-gamma in a murine breast cancer model. Asian Pac. J. Cancer Prev. 15.

Zacharakis, N., et al., 2018. Immune recognition of somatic mutations leading to complete durable regression in metastatic breast cancer. Nat. Med. 24.

Zeng, R., et al., 2005. Synergy of IL-21 and IL-15 in regulating CD8+ T cell expansion and function. J. Exp. Med. 201.

Zippel, D., et al., 2019. Tissue harvesting for adoptive tumor infiltrating lymphocyte therapy in metastatic melanoma. Anticancer Res 39.

Zhang, H., et al., 2005. Lymphopenia and interleukin-2 therapy alter homeostasis of CD4+ CD25+ regulatory T cells. Nat. Med. 11.

Zhang, L., et al., 2015. Tumor-infiltrating lymphocytes genetically engineered with an inducible gene encoding interleukin-12 for the immunotherapy of metastatic melanoma. Clin. Cancer Res. 21.

Zhang, B., et al., 2007. Induced sensitization of tumor stroma leads to eradication of established cancer by T cells. J. Exp. Med. 204.

Zhou, J., et al., 2005a. Telomere length of transferred lymphocytes correlates with in vivo persistence and tumor regression in melanoma patients receiving cell transfer therapy. J. Immunol. 175.

Zhou, J., Dudley, M.E., Rosenberg, S.A., Robbins, P.F., 2005b. Persistence of multiple tumor-specific T-cell clones is associated with complete tumor regression in a melanoma patient receiving adoptive cell transfer therapy. J. Immunother. 28.

Zhou, J., Dudley, M.E., Rosenberg, S.A., Robbins, P.F., 2004. Selective growth, in vitro and in vivo, of individual T cell clones from tumor-infiltrating lymphocytes obtained from patients with melanoma. J. Immunol. 173.